Safety Crimes

Crime and Society Series

Series editor: Hazel Croall

Published titles

Sex Crime (second edition), by Terry Thomas
Burglary, by R.I. Mawby
Armed Robbery, by Roger Matthews
Car Crime, by Claire Corbett
Street Crime, by Simon Hallsworth
Hate Crime, by Nathan Hall
Fraud, by Alan Doig
Murder, by Shani D'Cruze, Sandra Walklate and Samantha Pegg
Safety Crimes, by Steve Tombs and Dave Whyte

The royalties from this book will be donated to Hazards Magazine (http:/www.hazards.org/) and the Centre for Corporate Accountability (http://www.corporateaccountability.org/)

Safety Crimes

Steve Tombs and
Dave Whyte

WILLAN
PUBLISHING

Published by

Willan Publishing
Culmcott House
Mill Street, Uffculme
Cullompton, Devon
EX15 3AT, UK
Tel: +44(0)1884 840337
Fax: +44(0)1884 840251
e-mail: info@willanpublishing.co.uk
website: www.willanpublishing.co.uk

Published simultaneously in the USA and Canada by

Willan Publishing
c/o ISBS, 920 NE 58th Ave, Suite 300,
Portland, Oregon 97213-3786, USA
Tel: +001(0)503 287 3093
Fax: +001(0)503 208 8832
e-mail: info@isbs.com
website: www.isbs.com

First published 2007

Hardback
ISBN 978-1-84392-086-1

Paperback
ISBN 978-1-84392-085-4

British Library Cataloguing-in-Publication Data
A catalogue record for this book is available from the British Library

Project managed by Deer Park Productions, Tavistock, Devon
Typeset by Pantek Arts Ltd, Maidstone, Kent
Printed and bound by T.J. International Ltd, Padstow, Cornwall

Contents

Acknowledgements

We wish to record publicly our thanks to those who have offered unswerving support for the rather unfashionable criminological concern which forms the subject matter of this book, as well as for their constructive criticism. Joe Sim, Frank Pearce, Laureen Snider, Paddy Hillyard. Anne Alvesalo, Simon Pemberton, Roy Coleman, Jonny Burnett, Pete Gill, Courtney Davis, Gary Fooks, Erja Virta, Kas Wachala and Andrew Watterson are owed debts of gratitude for their advice and insights on material contained herein. We are privileged to think of all of these people as friends as well as colleagues. Our thanks also go to the many people who have attended the various meetings of the European Group for the Study of Deviance and Social Control, which has consistently given us space to air arguments expressed here; particular mention amongst these should go to Andrea Beckmann, for her friendship, and for her work in sustaining the 'British' section of the aforementioned organisation.

For his ready acceptance of the litany of promised, then missed, deadlines over the past years we are grateful for the support of Brian Willan for this project – in even taking on the text, he has done something which most publishers of criminology texts would not have. And much more than a nod of appreciation is also due to Hazel Croall who, as series editor, has read every word of this book, providing supporting yet detailed feedback, without which any of its current failings would have been far greater.

This book could not, quite literally, have been written without those people who devote their lives to struggling around those issues which we have only dared to attempt to describe and account for here. Notable amongst these in terms of both the extent of their contribution and their comradeship have been John Bamford, Hilda Palmer, Rory O'Neill, and Anne and Tim Jones. Special thanks go to David Bergman and to the (past and present) staff, board members, volunteers and supporters of the Centre for Corporate Accountability; the CCA's particular role within the struggle for safer workplaces – to provide the bereaved with a path through the tortuous bias and lack of acknowledgement of the criminal justice system – is a vital and humbling one.

As with any venture that spans a period of years, we could not have written this without the support, love and sacrifices on the part of those who share our lives. Lyn and Patrick, that's you. Kathrin, that's you. Thanks.

Finally, we'd like to put on record that what is here is the result of a working relationship and a friendship that has been, and will remain, invaluable to both of us. This text, for all its omissions and weaknesses, is much more than the sum of its parts.

List of tables and figures

Tables

Figures

List of abbreviations

ATP	Automatic Train Protection
BCS	British Crime Survey
CCA	Centre for Corporate Accountability
CPS	Crown Prosecution Service
DPP	Director of Public Prosecutions
DWP	Department of Work and Pensions
HASAW Act	Health and Safety at Work Act (1974)
HSC	Health and Safety Commission
HSE	Health and Safety Executive
ILO	International Labour Organisation
IMO	International Marine Organisation
LFS	Labour Force Survey
LOSC	Labour Only Sub-Contracting
MIC	Methyl Iso-Cyanate
ODPM	Office of the Deputy Prime Minister
OECD	Organisation for Economic Co-operation and Development
OSD	Offshore Safety Division
RoSPA	Royal Society for the Prevention of Accidents
SOLAS	Safety of Life at Sea
SME	Small- and Medium-sized Enterprise
TSR	Temporary Safe Refuge
UCC	Union Carbide Corporation
UCIL	Union Carbide of India Limited

Foreword

Rory O'Neill

The date 23 March 2005 was a business-as-usual day for BP, Britain's largest private firm. The board was feted as a model of corporate probity, and was delivering soaring profits and share prices. The global chief executive Lord Browne – the poster boy for caring, successful business – had even done the seemingly miraculous and given the petroleum giant a green tinge.

But a 'corporate blindspot' on safety left this safe bet for shareholders less safe for the workforce. It was 23 March 2005 when the Texas City refinery explosion exposed the deadly flaws in BP's boardroom strategy. Fifteen died and 170 were injured in a blast which razed large sections of the plant.

There was nothing unique about events at BP, except the company's size and profile guaranteed the disaster would perhaps command closer scrutiny from the media and authorities. One investigation report noted it had failed to learn the lessons of previous incidents, particularly a series of serious safety offences in 2000 at its Grangemouth plant in Scotland. Another reported it had slashed costs, all part of its 'cheque book mentality' towards safety. But at all times it had done what companies are first and foremost required to do – to be as profitable as possible.

This book is concerned with the nature of production and how this impacts on the decisions made in boardrooms. It also deals with how infrequently these decisions, however deadly, lead from the boardroom to the courtroom. It is not filled with stories of comic book villains with no concern for the consequences of their actions, portraying safety crimes as bad deeds by bad people. Lord Browne no more set out that morning to kill 15 workers than he set out to stop it. But boardroom decisions made with a key objective of maximising profit have consequences, and safety can be a casualty. Nice guys can kill you.

In fact, this book sets out to deal with only a fraction of this killing. Work-related health problems, not 'safety', have for decades been the bigger workplace killer. We should remember that jobs in service-sector Britain may for the current working generation become safer, but remain just as deadly through causing illness. More of us are in work than at any time in history.

Exposing the links between boardroom behaviour and deaths and injuries at work requires information spanning an unusually broad range of disciplines, from business to criminal law, industrial relations to industrial safety and economics to sociology. It may be this complexity that has meant that being hurt at work is not recognised as a crime of violence in Home Office statistics, despite the location of the crime being the only major feature distinguishing it from other acts of violence.

We don't know how many people are killed or injured as a result of work. We do know official figures are substantial underestimates. As the authors here establish, the majority of work-related deaths are excluded from the official statistics. The systematic exclusions include road traffic accidents in work time, deaths reportable to agencies other than the workplace safety watchdog the Health and Safety Executive (HSE) and local authorities, or deaths while working abroad or arising from exposures while working abroad. Tot these up and you reach a total in excess of 1,600 deaths a year – at least seven times the officially recorded level.

Examples of recent deaths at work that don't count in official occupational fatality statistics include the seven workers killed on 23 December 2006 when a helicopter carrying rig workers, all UK residents, crashed over the Morecambe Bay gas field. The deaths were investigated by the Civil Aviation Authority. The death of bus driver Michael Hallinan in December 2005, trapped between two buses, again escaped the work death records. It's the same story for Lloyd's Bank worker Lucy Wilson, 23, who contracted a liver infection while training call centre workers in India on a short work trip for the company, succumbing to the disease within three weeks of her return.

Even deaths that should be included by HSE's criteria frequently miss out, because the doctors charged with making the reports sometimes fail to see beyond the fatal injury to the deadly workplace.

Thus, even in terms of 'safety', the price is greater than it might at first seem. There are also health care and welfare costs, heartache and poverty for bereaved families, trauma for injured workers. Thomas Corr had almost all his right ear severed at the Luton IBC van factory while fixing a machine. Six years later, in May 2002, he took his own life aged 37 after suffering headaches, tinnitus and severe depression. His job killed him and devastated his family, but you won't see that reflected in any workplace fatality statistics.

This all has implications for prevention. If you don't count the bodies, the bodies don't count. When allocating resources and deciding public health priorities, over 1,600 workplace injury deaths each year will command a more serious response from government than HSE's artificially deflated headline figure of fewer than 250 worker deaths a year.

Social class defines those who will die or be injured. By HSE's own estimates, 'process, plant and machinery operators' are about ten times more likely to die or suffer a serious injury at work as 'managers and senior officials'.

Workers, though, are blamed. Not dissimilar to Union Carbide's response following the Bhopal gas leak, a report released by BP within two months of the March 2005 Texas City blast alleged it was 'surprising and deeply disturbing' mistakes by workers that led to the explosion. In response, the firm was accused of 'corporate scapegoating', a charge completely vindicated in subsequent reports which instead highlighted the culpability of the global BP board.

The idea of the accident-prone worker, a myth which has been examined and discredited so thoroughly the language should sound quaint and archaic, is, in the words of Canadian academic Bob Sass, 'a dead horse that won't lie down'. Blame-the-worker behavioural safety systems are growing in popularity, putting the focus on the individual worker rather than on the management-designed, operated, supervised and profit-oriented system that placed them at risk. Through threats and incentives workers are coerced into not reporting injuries. In the US, they call it 'blood-in-pocket syndrome' – don't report it, hide the injury until you get home.

Safety crimes are sanitised with the use of blame-free language – disasters, abuses, scandals. They are 'accidents', despite there being nothing accidental in the decisions and circumstances that caused them to occur. They are called anything but 'crimes'.

By contrast, there's plenty of language to suggest the problem is overstated or the creation of vested interests. We talk of 'ambulance chasers', but when factories explode there's no equivalent criticism of 'ambulance fillers'. When deaths occur, they are the product of 'freak' accidents, not the wholly predictable outcome of the way the job was designed and executed. Deaths in trenches or in falls aren't the everyday and obvious outcome of a management neglect of basic, simple safety procedures, but are 'one in a million' tragedies, apparently unforeseen and unstoppable. The police conclude, with unseemly haste and usually long before any form of credible investigation has taken place, 'there are no suspicious circumstances'.

As this rigorous examination of the evidence demonstrates, there is a long distance from the boardroom to the courtroom. Managers, directors and those appointing them accept without question there is a direct correlation between boardroom decisions and the company's performance. Large pay packets, bonuses and pensions can all be justified because there is a straight line from the guiding hand of the board to the productivity and profitability of the firm. But when things go wrong and workers die or are injured, then that guiding hand is not raised in an admission of guilt. It is hard to reconcile the two positions.

Of course, when it comes to dealing with legal liability, the boardroom does have certain advantages. For starters, company directors do not on the whole look like the popular perception of a criminal. Stereotypes point to yobs and yoofs and spivs, but the corporate criminals, however yobbish and anti-social their boardroom behaviour, just don't look the part.

It's not that these bosses are without fault. Investigations cited in this book suggest in at least two out of three workplace fatalities there is a clear criminal case to answer. But the higher you get up the corporate ladder, the more likely you are to have the cash and connections – actual, class and cultural – to play and win the legal game, as blame is 'mystified and obscured'. The criminal justice system denies almost all occupational injury victims and bereaved relatives an admission of culpability by the guilty parties or a day in court. Redress is normally only available through common law compensation claims, which, if they go to court, can provide an admission of negligence on a balance of probabilities (rather than a criminal offence). Common law compensation cases require companies to answer for their unsafe and unhealthy behaviour 40 times more frequently than HSE prosecutions.

HSE – the default workplace safety police – is far more likely to seek informal solutions and measures in response to clear evidence of a safety crime. HSE's strategy puts a new emphasis on voluntary measures, education and support, backed up by an argument that the real benchmark is improving health and safety performance. This is patently an important goal – although the evidence suggests there are few grounds for complacency on this count, and it cannot, in any case, be the only goal. Quiet encouragement ignores the role of deterrence. It also dismisses notions of justice and fairness.

Criminal breaches of safety law have to be viewed in the context of the drive by successive governments to lift regulatory burdens from business. HSE has adopted a business-friendly persona, which could reflect both directives from central government and the 'capture' of the organisation by business interests, who over time have skewed its outlook and work to be closer to that of the business world. And so all the major indicators of workplace safety enforcement action plummet: prosecutions, convictions, notices, inspector contact time and inspections.

Companies, meanwhile, opt into voluntary 'compliance' programmes, evading regulatory attention because their *reported* incident rates are low. These programmes are widespread in the US; government-encouraged variations on the theme are now appearing in the UK. Voluntary approaches have the clear potential to undermine regulation. And regulation is more than a benchmark against which errant companies can be judged and punished. It also includes the package of rights available to workers to raise concerns, to call for information and to demand action.

If safety regulation is weakened, either by deregulation or by an incremental shift taking more and more companies into the world of self-regulation, then these workers' rights could be weakened by stealth. This has created the space for companies, particularly large companies, to create their own wholly voluntary and self-enforced 'ethical' alternative to traditional enforcement. Corporate social responsibility and ethical trading are buzzwords in the boardroom. They are also great PR without the discomfort of external policing.

HSE has obediently embraced this model of voluntary compliance. BP, the firm with the 'corporate blindspot' on safety, whose global board was savaged in 2006 and 2007 in reports after the Texas City explosion and which was the recipient of Britain's biggest safety fine in 2002, was still featured in 2007 on HSE's corporate social responsibility webpages as an example of good 'director leadership'.

Yet companies themselves can be victims of under-enforcement. Good firms that invest in prevention and who give a greater priority to safety suffer a competitive disadvantage compared to those who only expend time and effort in the unlikely instance they are caught. It is a system that encourages the good to be not so good and the bad to stay bad.

When companies commit safety crimes they are not in their own terms necessarily dysfunctional; in fact, they are acquitting most of their duties exactly as their legal constitution requires. Companies are constituted to maximise profits from commercial activities, and failure to do this is an offence for which shareholders can and do demand redress. The system of corporate law and governance rewards profit and gives primary oversight to shareholders, a group whose principal concern is the size of those profits. Like muggers, corporations don't set out to hurt you, they just want the money and accept someone might, when things go wrong, get hurt in the process.

Their guilt, though, can be shockingly apparent. The fire and explosions at Piper Alpha, the sinking of the *Herald of Free Enterprise* and the death of student Simon Jones at Shoreham dock were, in the words of the authors of this book, 'all crimes that could be traced back to the carefully planned decisions of accountants and managers'. These crimes, they add, are 'committed in cold blood' and are often 'committed in order to advance financial goals or maintain profitable systems of production'. As US work organisation expert Charley Richardson of the University of Massachusetts Lowell puts it: 'They are killing us by diligence, not negligence.'

It is the legal 'controlling mind' doctrine that affords corporate muggers a near impermeable defence. The 'corporate personality' which faces safety charges enjoys all the rights of a real flesh-and-blood person, without similar responsibilities. This is useful to companies, as the corporate person has one clear defining trait – greed. Corporations are first and foremost about making money. Safety responsibilities, in legal and business if not moral terms, are a lesser concern. The duties are not absolute. The law says an employer must do what is 'reasonably practicable' to ensure work is safe and healthy. The law as it stands is not good at laying liability at the door of companies – it leaves too much 'reasonable' wriggle room and the volume and sometimes vigour of enforcement action is low. It is singularly ill-suited to making directors and top managers liable for safety crimes.

While there's little chance of being caught, there's little incentive to be law abiding. Inspection backed by enforcement is the most successful way to secure compliance. This effect would be amplified if fines were more closely related to the gravity of the offence or ability to pay.

Punishment, of course, is not a solution to crime, but it should be a possible consequence of it. It could deliver both a deterrent effect and, just as crucial, a recognition that lives at work are important, and harming working people is, like other acts of violence, a serious crime.

Recent experiences suggest it will take bigger, better targeted penalties to spur change at major companies. For BP, it took nearly two years, three critical reports, intense media scrutiny, a $21m fine, serious pressure from corporate responsibility lobbying groups and a general conclusion from critics that the reputation of the company and its golden chief executive Lord Browne was 'down the toilet' before it conceded in February 2007 that production and safety priorities had been in conflict in the run-up to the Texas City explosion, and that this was a problem that must now be addressed by better investment in safety and by less emphasis on production. Lord Browne said the Texas City blast 'was a watershed' that would 'forever change BP'. But two workers had died at the same plant just six months before. Prior to Texas City, BP already had more fatalities than its major US competitors. Deaths alone were not enough to change BP's mind.

Back in the world of 'real' crimes, bosses can and do spend time behind bars. In October 2006, Jeffrey Skilling, Enron's chief executive, was sentenced to 24 years in a US jail for corporate fraud. The collapse cost workers their jobs and pensions, and investors their fortunes. The same company had been responsible for deaths in the UK, but faced the usual fines. It was Jeff Skilling who flew into the UK after three workers died and one was seriously injured in an August 2001 explosion at the company's Teesside power plant. His comment then: 'We pledge to work with the authorities to ensure it never happens again anywhere.'

This wasn't the plant's first death or its first explosion. The company – not its executives – had been prosecuted successfully on two previous occasions, once after a worker was killed in the plant's construction in 1992, and once in 1998 after a worker received horrific burns in an explosion. There were all the signs of a dysfunctional company, but the book work looked OK and Enron claimed its finances were good, so no one looked too closely. Worker deaths don't hurt shareholders.

Directors remain effectively immune from safety law. An average of 11 a year faced court action for workplace safety crimes in the decade up to December 2005; just 11 were jailed. In 2005 alone, inspectors from the animal welfare body the RSPCA secured 31 prison or suspended sentences for animal cruelty. This was down from 44 in 2004. It investigated 110,841 complaints and secured 2,071 convictions of 1,031 defendants. On every count, the RSPCA massively outperformed the HSE, the human cruelty at work watchdog.

If there is a slow recognition that workplace injuries and fatalities are often the result of corporate safety crimes, this has not come about because of the emergence of new evidence. The evidence has always been there and the case has always been watertight. The reason this is on the agenda at all is because there has been a campaign to put it there. Bereaved families, trade unions, campaign and advocacy organisations like the Hazards Campaign, the Centre for Corporate Accountability, Disaster Action, Inquest, Families Against Corporate Killers (Fack), the Construction Safety Campaign and the Simon Jones Memorial Campaign have all called for legislation, for action and for accountability. They have an unanswerable case.

Despite technological advances and the decline in more obviously hazardous industries like manufacturing and mining, some new trends in modern production are placing a large slice of the workforce at greater risk. More of us are now contingent workers – temporary and insecure, with few rights and little scope to object to safety abuses. The benefits system is getting meaner and sickness systems more punitive, so the pressure to stay in work, however hazardous, could be increasing. Working longer, harder and faster are not hazards dealt with under the current enforcement model, but they are the emerging problems of the twenty-first century workplace.

Workplace health and safety isn't a battle won; it is barely a battle started.

Rory O'Neill
Hazards magazine (*http://www.hazards.org*)
May 2007

Chapter 1

Introducing safety crimes

Introduction

If you were to stop one hundred people in the street and ask them to list, in their opinion, the five most serious crimes in Britain today, it's a very safe bet that in the vast majority of responses no form of corporate crime would feature. This is not to be disparaging of popular understandings of crime, for these ways of seeing are socially constructed. So, check today's newspapers for reports of political discussions or Home Office statements on law and order: again, corporate crime will be absent. Now scour those newspapers more thoroughly, and try to find corporate crime stories – corporate crime may be covered, but is unlikely to be given prominence, nor discussed in the language reserved for 'real' crime, and is more likely to be financial 'scandals', in specialist sections far from the headline-grabbing news pages. Check your TV guide for today's viewing, and search for corporate crime coverage amongst what will no doubt be a great deal of fictional, real-life and documentary programmes on crime – and it would be a surprise to find much, if any.

Now if corporate crime – in essence, crimes by corporations in pursuit of their legitimate goals[1] – is relatively invisible, then this invisibility is even more marked for some of the subcategories covered by this general term. One such sub-category – safety crimes[2] – is the subject of this text. Safety crimes – which we define as violations of law by employers that either do, or have the potential to, cause sudden death or injury as a result of work-related activities – are, we shall argue throughout this book, ubiquitous, have devastating physical (and psychological) effects, and carry with them enormous financial and social consequences. Despite this, such crimes remain relatively invisible at the level of political and popular consciousness. Indeed, where events that *may* be represented as safety crimes do emerge in such contexts, they tend not to be represented as such: instead, they are likely to be viewed as accidents, disasters, tragedies, and so on. Even less forgivably, safety crimes as an area of focus is also obscured by the overwhelming focus of criminological teaching, research and writing.

It is this relative invisibility that is the central rationale for this text. We are concerned here with how this is achieved, why this is the case, and with the consequences of such crimes being obscured. Further, we wish to examine the role played by institutions and processes in maintaining definitions of crime from which safety crimes are marginalised. And, finally, we seek to set out some of the conditions for overcoming this relative invisibility. That is, we ask in what ways might governments, the media, the law and criminal justice system, and, indeed, criminologists, take safety crimes more seriously – and what might the effects be of any emergent focus upon this widespread area of offending.

If the invisibility of safety crimes is a central, recurrent theme of this book, then perhaps our first task is to put some safety crimes before our readers' eyes: what are safety crimes, what do they look like, what forms do they take, where do they occur, what consequences do they have, who are their victims, and how does law and its agencies respond to them? For this reason, the bulk of this chapter is devoted to a series of brief case studies of such crimes, to establish contours of the all-too-unfamiliar phenomenon with which the book is dealing. Prior to turning to these cases, we seek briefly to mark out the terrain of the book through a definition of safety crimes.

What are safety crimes?

As we have indicated above, safety crimes, as used throughout this text, are a subset of corporate crimes. And our understanding of the latter follows what has by now become a fairly standard definition, based upon the work of Kramer (1984), Clinard and Yeager (1980: 16), Schrager and Short (1978), and Box (1983). Thus, by 'corporate crime', we mean

> Illegal acts or omissions, punishable by the state under administrative, civil or criminal law which are the result of deliberate decision making or culpable negligence within a legitimate formal organisation. These acts or omisions are based in legitimate, formal, business organisations, made in accordance with the normative goals, standard operating procedures, and/or cultural norms of the organization, and are intended to benefit the corporation itself. (Pearce and Tombs 1998, 107–110)

We can, for the purposes of this text, simply replace corporate crime with safety crime in this definition. And this definition has a series of useful characteristics, to which it is worth drawing immediate attention.

Safety crimes may constitute violations of the Health and Safety at Work Act 1974, the primary piece of safety legislation in the UK, which is a criminal statute. Equally, however, there is legislation governing occupational safety that falls beyond the criminal law. Within the above definition, then, through reference to illegal rather than criminal acts, it avoids the restriction that follows from the fact that many laws, enforced by administrative bodies through the civil courts, also regulate actions which cause injuries to specific individuals or which undermine social institutions. Indeed, following Sutherland, the content of laws and the nature of legal distinctions such as those between crimes, torts and administrative sanctions, between acts *mala in se* and *mala prohibita*, are conventional, time-bound social products without an intrinsic substantive meaning that transcends their social or historical contexts; what acts or omissions constitute crime must be understood in terms of contingency (Lacey 1995). Thus, we extend our definition of crime beyond that proscribed by criminal law or indeed that which has been processed through the legal system.

Second, through references to negligence and the use of the phrase 'omission or commission', the definition avoids the trap of arguing that for corporate crime in general – and safety crime in particular – to exist there must be *actus rea* and *mens rea* ('knowing mind'). Given the organisational locus or origins of safety crime, to emphasise either an illegal *act* or an *intention* is often inappropriate, for two reasons. First, each term is anthropomorphic and individualising, and by definition is problematic in its application to a corporate entity which is more than, or different to, the sum of individual human actors. Second, intention implies a relatively unproblematic link between an act or omission and its consequence – yet this simplistic causal sequencing leads to an obscuring of the construction and maintenance of a situation or context which, as a consequence, is fertile ground for violations. Any focus upon safety crimes requires us to examine these in terms of their *organisational production*.

On the former point, it has been argued that 'most corporate crimes cannot be explained by the perverse personalities of their perpetrators' (Braithwaite 1984: 2), and this claim calls into question the proclivity within individualistic liberal or bourgeois cultures to locate the source of evil deeds in evil people (*ibid.*; see also Shrager and Short 1977: 410, Snider 1993: 61). Corporate and safety crimes can be produced by an organisation's structure, its culture, its unquestioned assumptions, its very modus operandi, and so on. Thus to understand such phenomena must not obscure human agency, but does require a shift from abstracted, atomised individuals to account for agency in the context of structures.

The definition of safety crime used in this text is, then, an explicitly inclusive one. That is, throughout this text we discuss *as crimes* many acts and omissions which have not been subject to any formal judicial process; thus these are phenomena which have not been interrogated as, let alone proven to be, violations of the criminal, or indeed any other form of law. To fail to do otherwise would lead us into the study of what Braithwaite called class-based administration of criminal justice, thereby obscuring precisely the most useful insights of Sutherland regarding the class-biased development of law and its differential implementation (see Chapter 4).

Other issues regarding our definition of safety crimes, and thus the scope of this text, require brief clarification.

First, we should emphasise that we focus here upon safety rather than health. While the commonly-used phrase 'health and safety', and the very object of legislation in this area, often reinforces the intimately linked nature of these two concepts, consideration of each raises some common, but many distinct, phenomena. Violations against worker or public safety are – usually – immediately apparent through near-miss, injury or even death; violations of occupational health law often involve a much more complex and contestable causal chain, the key factor often being a state of affairs rather than an event, such as long-term exposure to a noxious substance or endurance of a unhealthy mode of working which can produce cancer, stress or arthritis much later in one's life. This causal complexity makes the burden of proof a difficult one for victims of such crimes – involving, not least, often confounding resort to complex and often contradictory scientific claims. Similarly, the immediate visibility of safety offences as opposed to health offences means that it is the former that are much more likely to result in formal enforcement action on the part of regulators – rare though any form of such action is. Partly related to this is the fact that if occupational injury data is incomplete and problematic, to say the least, as we shall see in Chapter 2, that relating to occupational ill-health is almost non-utilisable – problems of measurement beset the area of occupational health to a far greater extent than that of occupational safety. And, again perhaps relatedly, there is also a sense in which meeting legal duties to provide a safe working environment imposes less costs in terms of changes to work organisation, practices and so on than dealing with less tangible ill-health effects.

None of this is to relegate occupational health crimes below that of safety crimes in social importance. But it is to say that there are important differences, so that an adequate treatment of each would require a much longer text than can be presented here. Further, none of this should imply that there are not common causes of occupational safety and health crimes – each can be understood, for example, through many of the causal factors or characteristics of acute deaths as set out in the table at the end of this chapter.

Second, we should clarify that while our focus is upon safety crimes, neither the legal duties that are being violated by these, nor the effects of such violations, are restricted to workers. Thus, for example, the Health and Safety at Work Act 1974 extends the duty of care owed by employers not simply to employees and the self-employed, but also to the general public as well as others, such as sub-contractors, not in direct employment but likely to be affected by the conduct of the employer's undertaking (sections 3–5). This is a recognition that whilst the victims of safety crimes are often employees, no actual nor metaphorical factory fence can exist that protects members of the public from decisions and omissions within companies that produce offences with actual or potentially dangerous effects. And the force of this legal requirement is under-scored in some of the cases selected for consideration below. Thus, for example, while the gas leak at Bhopal killed employees at the Union Carbide plant, arguably far greater devastation was wrought upon the local community, where tens of thousands were exposed to the gas, and to the surrounding natural environment, much of which remains despoiled to this day; the victims of the failure to ensure a safe system of work aboard the *Herald of Free Enterprise* were mostly consumers, that is, passengers; and the four deaths which arose as a result of the faults of Transco to work safely were of a family to whom the company supplied gas. Business organisations supply goods or services; those to whom those goods or services are supplied, or who live in the proximity of areas where some of that production or supply is undertaken, are just as likely to be victims of safety crimes as workers.

Finally, it remains here to be stated that throughout this text we view safety crimes as crimes of *violence*. Though this is not an original approach, it is, as we shall see, one which, in the context of academic criminology and general representations of occupational injuries, a rarity. For us, this lack of criminological attention to safety crimes as crimes of violence is less a quality of the latter phenomena, more a failure of the discipline to reflect upon long-standing, but ontologically weak, assumptions. Thus once occupational injuries are viewed not as accidents but as incidents which are not only largely preventable, but which the law requires to be prevented, then they fall within the ambit of criminology. Then, if we consider these illegalities in terms of their potential or actual consequences – injury and death – we realise that these look remarkably similar to the results of those events that most men and women, as well as policy-makers, politicians, and academics, deem to be 'proper' violence. Most crucially, these latter conceptions of violence are, as we shall see, based upon an implicit association of violence with the inter-personal and with intention, both qualities that are often absent from safety crimes.

That safety crimes are crimes of (actual or potential) violence is, for us, unequivocal; yet to reach this conclusion one must generally move beyond criminology, to understandings of violence developed in other disciplines.

One such holistic understanding of violence has been developed by Jamil Salmi who, following Galtung amongst others, has set out a systematic analytical framework which aims to explore a variety of dimensions of this phenomenon; he then identifies the patterns and relationships linking the various manifestations of violence to the prevailing economic, social and political power structures in an effort to establish accountability. Salmi sets out four broad categories of violence, and indicates a series of type within each category.

First, he identifies 'Direct Violence', that is, deliberate injury to the integrity of human life, which includes homicide, brutal acts and restrictions or physical constraints. It is within, though not encompassing all of, this category of violence that the overwhelming majority of criminological research proceeds. Highlighting intent through reference to *deliberate injury* (Salmi 2004: 56), this would exclude most forms of deaths and injuries associated with working.

Second, Salmi notes 'Indirect Violence', namely indirect violations of rights to survival; this includes the sub-categories of violence by omission, which 'draws on the legal notion of non-assistance to persons in danger' (ibid. 57), and mediated violence, 'which is the result of a deliberate human intervention in the natural or social environment whose harmful effects are indirect and often delayed' (ibid. 58). In the context of a discussion of the latter, Salmi considers the mediated violence generated through the capitalist production process, noting the tendency towards health and safety effects which is an effect of the inherent desire of employers to minimise 'unproductive' expenditures (Salmi 1993: 56).

Third, he refers to 'Repressive Violence', that is the 'Deprivation of Fundamental Rights' which includes violations of social, civil and political rights. This would incorporate as studies of violence much of the burgeoning literature around human rights and their violations – a literature dedicated to considering issues which if 'not completely ignored by criminology as a discipline' remains 'a minority pursuit', and for the most part a very recent one at that (Carrabine *et al.*, 2004: 352).

Fourth, and finally, Salmi identifies 'Alienating Violence' – 'Deprivation of "Higher" Rights' – based on 'the assumption that a person's well-being does not come exclusively from fulfilling material needs, the notion of alienating violence refers to denying a person the right to psychological, emotional, cultural, or intellectual integrity' (2004). This includes alienating living and working conditions, social ostracism and ethnocide (Salmi 1993, following Galtung 1981).

Categories of 'indirect violence' and 'alienating violence' are particularly important in thinking about safety crimes, as they direct our attention to the 'relations of domination, and the violences that are condoned as part of the "normal" and "healthy" functioning of a society such as ours' (Catley 2003: 5). Safety crimes, as we will argue at various points in this text, result from the organisation of production and working regimes that carry deadly consequences for some workers and members of the public. It is this that clearly puts safety crimes on the terrain of 'indirect violence', since safety crimes are the human consequences of mediated violence, inflicted indirectly through working. The concept of 'alienating violence' is relevant for safety crimes since it allows us to move from the level of the individual to *structural* violence, that is, violence that is produced systemically or structurally as a consequence of a variety of key organisational or social institutional arrangements. Key institutions that produce violence structurally and systemically in the context of work include labour markets, the occupational structure and the labour process, the uneven distribution of power within and around work organisations, the corporate form *per se*, and constituent parts of the state, such as criminal justice and legal systems, regulatory bodies and Government ministries. At various points throughout this text, we will therefore consider how those organisations and institutions play a role in mediating the conditions of indirect violence, and in the structural production of violence. This is how we approach the case studies that follow.

Case studies in safety crime

Introduction: using case studies

The case study approach has been, and remains, particularly important in all forms of corporate crime research. There are several reasons for this general tendency to produce case studies. First, in almost all jurisdictions there is a lack of consolidated corporate crime statistics that form basic raw material with which corporate crime researchers might work. Further, projects designed to generate large scale quantitative data with which then to work are extremely expensive and thus require funding – but such funding is unlikely to come from either governments or employers, each of which are likely to find corporate crime research unattractive; such funding has thus been extremely rare, is now virtually non-existent (Snider 2003), and there have been very few such large scale surveys (see Slapper and Tombs 1999: 37–41).

Thus there is a tendency in corporate crime research towards studies of discrete crimes or groups of crimes. Typically, these crimes have been large-scale in nature, that is, involving significant amounts of financial loss, or large numbers of victims (most commonly, consumers, employees, or local residents), or with particularly symbolic political or popular effects. These studies go beyond stating the 'facts' of particular cases, but generally involve detailed *post hoc* social scientific reconstruction of events and processes surrounding a particular crime or group of crimes; somewhat differently, some of these studies are based upon intimate, after-the-event accounts by whistle-blowers or investigative journalists. The result is generally a detailed case history which facilitates understanding of the processes by which corporate crimes emerge, occur, and are responded to. Albeit presented in truncated and outline form, this is precisely one of the tasks of the cases here.

Case study research within corporate crime does have advantages. First, a series of discrete case studies can provide an overall insight into the enormous *range* of offences in which corporations and organisations can and do become implicated, from the well-known to the obscure, from the 'trivial' to the more obviously serious. Second, taken together, this work reinforces the fact that business offending results from almost every business activity, in almost every area of economic activity, amongst corporations and organisations of all sizes. Third, a series of case studies, each of which is ostensibly quite different in terms of industrial and geographical location, the key actors involved, the scale and nature of consequences, and so on, may generate recurrent themes from which at least tentative generalisations can be generated. As Braithwaite has argued, case studies – producing a 'qualitative understanding of the contours of corporate crimes and how they unfold' – are 'most likely to advance our understanding of corporate crime as a social phenomenon' (Braithwaite 1984: 8, 7; see also Coleman 1987: 429).

The construction of case studies of offences (and of types of offence or offending industries), often involves trawling for evidence and source material beyond the confines of the discipline of criminology. Research on, and evidence concerning, corporate crime, needs to be collated from a range of disciplines which have no explicit concern with either of what Garland has called the Lombrosian or governmental projects, and out of which contemporary criminology has emerged (Garland 1994). Thus the sources upon which we draw in constructing our cases, and indeed which we use more generally throughout this text, are not primarily criminological, indeed criminological sources are often in little evidence here. Studying safety crime means moving well beyond criminology, drawing upon literature in business, management and organisational studies, economics, history, political economy, politics, and sociology, and which makes reference to substantive areas of study such as industrial relations, business ethics and/or studies in the social responsibility of business, studies of regulation, science–technology policy, studies of corporate failures, disasters and crises, and studies of social movements and activist politics.

Of course, care must be taken in interpreting cases, since, as we have already noted here, using and working with case studies, particularly in the context of safety crimes, raises issues of reliability. Thus it is important to note that in contrast with other methods used in the social sciences, case studies rely upon sources of evidence commonly derived from investigative journalists and whistleblowers rather than sources that can more easily be represented as 'hard' academic research (although, see George and Bennett 2005). Dependency upon investigative sources and whistleblowing therefore invites charges of politicisation, polemics, and moralising (for a somewhat scathing attack around precisely such points, see Shapiro 1983). Moreover, each of the events described of course raise specific issues that relate to very specific, contingent, circumstances. Having said this, in several of the cases highlighted, the failure to heed the warnings of whistleblowers and concerned members of the public were key features of the circumstances leading to the crime. Often investigators' and whistleblowers' voices are ignored until after the event. Thus, although we have to recognise that those voices may not be completely impartial, this is no reason to continue to ignore them. Indeed, the great tragedy is that it often takes a fatal incident to sit up and take whistleblowers and other marginalised sources of warning seriously.

Conclusions drawn from any particular case study may not necessarily apply beyond the specifics of the case; 'the use of a case study can be diminished by a belief that the findings may be idiosyncratic' (Bryman 1988: 88). The cases (at least those of *incidents*) presented here are *atypical*: most safety crimes are mundane rather than exceptional, do not result in deaths, let alone multiple fatalities; most are not subject to any investigation, let alone legal proceedings; most do not register at all in terms of publicity. These very characteristics combine to make such incidents, which occur every day up and down the UK and, in fact, across the globe, simply invisible and thus not amenable to study. The fact that most of the cases presented here are not typical therefore does not undermine their usefulness as case studies, for, as we have argued above, to present a case study is not to claim representativeness. Further, to echo Braithwaite, above, there is still much to learn from surveying these cases. It is important, therefore, not to draw general rules cast in stone from cases, but rather to identify common themes and features of cases to enable us to open up questions about the circumstances of causation which might shed some light upon how we might achieve more effective prevention. This at the very least makes them social facts that are worthy of critical attention.

Why these as opposed to other cases? Several criteria informed our choice of cases, but two are worth highlighting. First, to construct a case requires that one knows about it and that one can find out more about it – in other words, it requires a level of visibility that most safety crimes, mundane and routine as these are, simply never achieve. Notwithstanding this obvious but important limitation for presenting a range of case studies, we have sought here to include the most infamous and internationally-recognised cases, alongside some which caused barely a ripple beyond the immediate vicinity within which they occurred, or the group of people most immediately affected. Second, taken collectively, we have sought to illustrate safety crimes in all their variety. Thus, in the following, there are cases where one or several workers are the victims, and others where the victims are or include consumers, local communities and/or the natural environment. Third, the cases illustrate a range of organisations as offenders – from the large multinational conglomerate through to smaller companies to those operating at the borders of the legal/illegal economy. Fourth, we have sought to include cases where legality arises intentionally alongside those where the key issue appears to be that of negligence. Fifth, and finally, an industry focus was taken, in the case of the construction industry, to highlight how the structure and culture of a sector can be, at the very least, crime-facilitative.

Our first case study, then, focuses upon an industry which appears potentially criminogenic, at least if occupational deaths and injuries are an index of safety offences, namely the construction industry. Six further cases focus upon discrete 'events', setting out briefly the circumstances surrounding: the gas leak at the Union Carbide Chemical plant in Bhopal, India; the capsizing of the cross-Channel passenger ferry the *Herald of Free Enterprise*; the fire and explosion on the Piper Alpha oil installation; the death of Simon Jones; the gas explosion in Larkhall, Scotland that killed a family of four; and the deaths of 23 Chinese cockle pickers at Morecambe Bay. And a final case study focuses on an ongoing series of safety violations at the Sonae Chipboard factory in Kirkby, Merseyside. In sum, these cases represent a range of incidents and contexts, the scale of deaths and injuries produced, their visibility, their industry, and even their geographical location. All have been subject to some kind of legal process which places them firmly within the principal concern of this book: safety crimes.

UK construction

Key causal issues highlighted in this case include: casualised, sub-contracted and increasingly migrant workforce; long and complex supply chains; legal/illegal business relationships; aggressive management; market pressures; industry norms; problems in regulatory processes (see Table 1).

As the Health and Safety Executive's[3] most recent annual report notes, the construction industry is probably the last remaining heavy industry of any size in Britain. It employs 2.2 million people, and contributes up to 10 per cent of GDP, making it the single biggest industry. And it is a growing industry with significant labour turnover which, according to the Construction Industry Training Board, 'needs at least 87,000 recruits a year' (Construction Skills Network, 2006). But it is also rather a peculiar sector – and these peculiarities tend to combine to make its workforce particularly vulnerable.

The industry is characterised by a large number of companies, most of them small, which are linked through complex systems of sub-contracting and long supply chains. Eighty-five per cent of the work in the industry is done by supply chain organisations, the majority of which are regional small- and medium-sized enterprises (SMEs). This makes locating responsibility and accountability, training, and unionisation all very problematic issues. Self-employment is significant in the sector, and it is growing, given the spread of 'Labour Only Sub-Contracting' (Anderson and Rogaly 2005: 33, and passim). High levels of sub-contracting, agency workers, LOSC (Labour Only Sub-Contracting) and self-employment come together to give the construction industry the highest proportion of full-time self-employed workers of any industry in the UK (London Assembly Health and Public Services Committee 2005: 10).

There are substantial numbers of migrant workers from the EU accession countries working in the construction industry in London, especially workers from Poland and the Czech Republic. New arrivals might well have difficulty understanding health and safety briefings reporting processes or their rights at work.[4] Employers may also employ them illegally, both as a means of reducing costs and as a way of driving down labour conditions in general on a site or a job. More generally, the industry has a highly mobile workforce. Construction workers tend to move from project to project and this may mean that injuries often go unreported, while it also means workers receive little or no safety training (ibid.).

There is a large informal construction sector, with some £4.5 billion to £10 billion construction work nationally undertaken 'cash in hand'. Companies who do not pay their taxes are also likely to have a 'less safe working environment' for their workers; further, the informal economy creates market conditions that put 'pressure on legitimate builders to cut corners in order to compete for work' (ibid.: 10). It is, therefore, a highly casualised industry, and casualisation is a major source of danger for workers: according to the Health and Safety Commission,[5] the annual injury rate to workers with short job tenure (less than six months) is 5.7 times that for workers whose job tenure is at least five years (Health and Safety

Commission, 2003a: 3), while over one in five of all reportable injuries are sustained by workers who have been with an employer for less than a year (ibid.). Further, we know that the risk of death and injury is far greater for self-employed workers (see Chapter 2). Little wonder, given the above, that the HSE has characterised the national construction industry in the following way: 'no entry threshold, highly fragmented, itinerant and casualised' (London Assembly Health and Public Services Committee 2005: 10).

Little wonder, too, that construction is also one of the UK's most dangerous sectors. The sector has a fatal injury rate of over five times the all-industry average, and is the sector with the highest number of deaths (Health and Safety Commission 2005a: 13).[6] The number of workers killed in the construction sector each year in Britain has remained in the region of around 70–80 since 1996/97. In London the construction sector employs 5 per cent of London's workers, but accounts for 50 per cent of fatal injuries and nearly 20 per cent of major injuries in the city's workplaces' (London Assembly Health and Public Services Committee 2005: 1). Moreover, the high level of self-employment drastically undermines reporting rates – by up to 95 per cent. Yet even the reported figures, if a vast under-estimate, obscure even greater levels of risk for some workers within the sector. As the Southern and Eastern Regional Trades Union Congress (2005) has recently noted, 'What must also be taken into account is that up to a third of workers within the sector are employed in office and desk-based work and, as such, are not subject to construction-specific hazards. As such, accident rates as a percentage of the total construction workforce should be considered with this in mind'. Thus the rate of fatal and serious injury varies significantly between occupations within the industry. The most dangerous trades include steel erectors (41.2), crane drivers (32.8), roofers and cladders (24.6) and scaffolders/steeple-jacks (21.2).[7]

The main causes of fatal injuires are: falling through fragile roofs and rooflights; falling from ladders, scaffolds and other work places; being struck by excavators, lift trucks or dumpers; being struck by falling loads and equipment; and being crushed by collapsing structures (www.hse.gov.uk/construction). In other words, these are incidents with mundane, and highly preventable, causes. As the employers' body, the Federation of Master Builders, put it, 'there is nothing intrinsic about construction that suggests that somebody has to die … [yet the] culture has become so engrained in that construction is dangerous, therefore someone is always going to get hurt' (cited in London Assembly Health and Public Services Committee 2005: 10–11).

While we would resist arguments which reduced the levels of injury and death in any industry to 'culture' (see Beck and Woolfson 1999), what the above does point to is that construction is not intrinsically dangerous, rather the source of risk for workers being found in the organisation of the sector and in work therein. These are socially constructed, and can and do vary between time and place. A similar point has been made about the very different 'safety' records of the Norwegian and UK offshore oil industries, a very useful comparison since it spans effectively the same work in the same context, that is, the North Sea which, whilst clearly inhospitable, is not inherently dangerous in the sense that it necessarily produces high numbers of worker deaths and injuries. Thus Ryggvik (2000) suggests that improved offshore safety in Norway is particularly the result of rights for union representatives to stop work when they feel safety is jeopardised, as well as the maintenance of strong offshore unions with a comprehensive network of trade union-appointed safety representatives; this is in marked contrast to the strident anti-trades unionism of the UK sector (Whyte 1999a).

The very scale of the industry (both in terms of numbers employed and worksites), the high level of sub-contracting and self-employment, the relatively short-term nature of many construction projects, and the scale of the informal economy make the work of the HSE particularly difficult in this sector (London Assembly Health and Public Services Committee 2005). Research using HSE internal data for the years 1996–2001 found a 52 per cent decline in the numbers of inspection contacts across the sector (Unison/Centre for Corporate Accountability 2002). Notices and prosecutions are in decline across the sector. Thus data for the most recent year, 2003/04–2004/2005, shows a decline in all formal enforcement activity in the construction industry: from 3,487 notices in 2003/04 to 2,481 in 2004/05 (from 798 to 548 improvement notices, from 33 to 20 deferred prohibition notices, and from 2,656 to 1913 immediate prohibition notices), an overall fall of almost 40 per cent (adapted from www.hse.gov.uk/statistics /enforce/index-ld.htm). Data for the same years also shows a fall in offences prosecuted in the industry (from 617 to 550 in 2004/05) and in convictions secured (from 418 in 2003/04 to 396 in 2004/05).[8]

Yet these trends fly in the face of evidence of the scale of offending across the construction industry: the subject of frequent enforcement 'blitzes' by the HSE, one such blitz in March 2005, for example, saw 214 enforcement notices issued and work stopped on 244 sites in the course of 1,170 visits which had been heavily publicised in advance (Trades Union Congress 2005a). In October 2005, targeted inspections of 1,379 contractors, focusing on falls from low height in the industry, led to 134 prohibition notices and 36 improvement notices being issued related to working at height (www.hse.gov.uk/construction/fitout). A 2003 blitz, involving 1,429 site visits, resulted in 332 prohibition and 82 improvement notices being served. Thirteen potential prosecutions were also noted. As in a blitz earlier in the same year (June 2003), HSE concluded that 'Once again, over a third of construction sites were well below standard' (Health and Safety Executive 2003a).

Rather indicating a high level of safety crime across the sector, a recent report upon the industry, drawn up having taken evidence from a range of interested parties, has noted the existence of a 'combative culture' within construction:

> Several organisations identified a 'competitive combative culture' within the industry, in which the cause of safety problems is obscured as companies seek to avoid financial or legal penalties. This culture tends to reduce reporting of injuries both on sites and to HSE. Others noted a persistent complacency about worker injury. (cited in London Assembly Health and Public Services Committee 2005: 10)

There are good reasons, then, for thinking that the construction industry is criminogenic.

Bhopal, December 1984

Key causal issues highlighted in this case include: power and reach of multinational corporations; 'developed' versus 'developing' economy safety standards; poor design, maintenance, and management; global market pressures; victim blaming; power to circumvent/prevent legal processes; safety and environmental offences; long-term victimisation – lack of compensation (see Table 1).

In Bhopal, India, a chemical plant, operated by Union Carbide of India Limited (UCIL), a subsidiary of Union Carbide Corporation (UCC), used highly toxic chemicals, including methyl iso-cyanate (MIC), to produce pesticides. On the night of Sunday December 2, 1984, water entered an MIC storage tank setting in process an exothermic reaction. Soon, a cocktail of poisonous gases, vapours and liquids, including up to forty tons of MIC, was spewing into the atmosphere.[9]

The Indian Government initially put the number of acute deaths at 1,700, a figure subsequently revised to 3,329. Twenty years after the leak, in 2004, Amnesty International estimated that there had been over 7,000 such deaths, with 15,000 people having since died from longer-term effects. About 100,000 'survivors' will never work again. These 'human' effects do not begin to account for the environmental damage caused by the leak – and still being caused, since Dow Chemical, the world's largest chemical company which effectively took over UCC in 2001, refuses to accept any responsibility for 'cleaning up' the affected surroundings.

Union Carbide Corporation worked hard at influencing both public opinion and the legal process through a series of largely spurious arguments about the incident, the nature of the Indian company, and the Bhopal plant and its employees. These included claims that: Bhopal's safety standards were identical to the standards at UCC's plant in Institute, West Virginia; it had an excellent safety record, and the design of the plant's Standard Operating Procedures – UCC's responsibility – was basically sound; the production of MIC in India, the siting of the plant and the quality of the materials used, were all the responsibility of UCIL and the Indian State; UCIL was an independent company responsible for its own affairs; India's 'cultural backwardness' was responsible for the poor maintenance and management, poor planning procedures and the inadequate enforcement of safety regulations; and, later, that the accident was due to sabotage, on the part of 'Sikh extremists' and then an errant tea-boy. These are classic strategies in the naming of an event as an 'industrial accident'.

When the leak occurred, key safety features were either inoperable or inadequate to their task: the vent gas scrubber was turned off, the flare tower was inoperative, and hoses that might have doused the gas had insufficient water pressure to reach the stack from which it was escaping. Further, while the storage tanks should have been refrigerated, the refrigeration unit had been turned off to save $50 per week. There are also serious questions that need to be addressed regarding the plant design itself. Plant instrumentation was inadequate to monitor normal plant processes. The refrigeration plant at Bhopal, even when working, was not powerful enough to cool all the MIC stored there, and the vent gas scrubber and flare tower were only designed to deal with a limited type of emissions.

In these respects, Bhopal was demonstrably inadequate and inferior to Institute. Nevertheless, even with this inferior technology, far fewer people would have died if: the plant had not been sited near shanty towns; there had been adequate risk assessment, modelling and monitoring of discharges, and emergency planning and management; plant personnel, local medical services and the state and national government had known more about the nature and effects of the deadly gaseous emissions – in the immediate aftermath of the leak, UCC refused to divulge any information about the chemical content of the leaked substances, making it very difficult to properly treat victims.

Further, UCC clearly owned and controlled UCIL; UCC owned 50.9 per cent of UCIL and exercised significant control over it. UCIL's production and marketing strategies were dictated by the corporate strategies of UCC; UCC had dictated how and which chemicals were produced and stored; UCC monitored safety procedures and UCIL was forced to rely upon UCC for technological assistance and updates. Indeed, UCIL's production of the pesticides Temik and Sevin took place under commodified conditions: they were to be produced and sold in such a way that sub-divisions of the company showed a normal profitable return on investment. It is questionable whether it was possible for UCIL to do so safely. And market conditions were changing: following two decades of huge growth, the pesticides market in India had become extremely competitive by the end of the 1970s, and by the beginning of the 1980s pesticide demand in India had all but collapsed. Thus the industry became characterised by harsher and increasing levels of competition.

What is clear is that UCC seems to have been responsible for both the acts of commission and omission that created the Bhopal disaster. The contentions made by UCC concerning the Bhopal disaster in its publicity and its legal arguments do not stand up to scrutiny.

On visiting Bhopal in the aftermath of the leak, Warren Anderson, UCC's CEO, was arrested – but released days later, never to return to India. Although, in December 1991, the Chief Magistrate in Bhopal ordered Anderson and the company to appear in court on charges of culpable homicide, neither appeared and are listed as 'proclaimed absconders'. Overwhelmingly, legal activity centred on civil suits for compensation – and, following a flurry of such suits, the Indian Government assumed legal powers to secure damages in March 1985 on behalf of victims who were not consulted. The initial sum demanded by the Indian government was $3.3 billion; in 1989, UCC and the Indian government reached an out-of-court settlement of $470 million. This rendered UCC immune from all litigation, including criminal charges. The money was to compensate the families of the 3,329 deaths and 20,000 seriously injured officially recognised by the Indian government.

But the money has still has not been fully distributed. Union Carbide, now owned by Dow Chemical, paid out £250m in compensation to residents in 1989, but only a part of that sum has been distributed. In July 2004, India's Supreme Court ordered the government to distribute money held in the bank, currently worth £174m, to the 566,876 Bhopal survivors and relatives whose claims have been successfully lodged. As late as September 2004, around US$330 million remained held by the Reserve Bank of India.

The Herald of Free Enterprise, March 1987

Key causal issues highlighted in this case include: market pressures; industry norms; systematic failure to heed warnings; poor design, maintenance, and management; victim blaming; and failings in legal and regulatory processes (see Table 1).

The *Herald of Free Enterprise* was one of the largest roll-on, roll-off (ro-ro) sea ferries – so-called because vehicles drive in through the bow and out through the stern at the destination. But when using the Belgian port of Zeebrugge, where turnaround time was longer than on other routes due to the design of loading facilities, it was commonplace to leave the port with the bow doors open – even though it was the formal responsibility of the assistant bosun to see them closed (Boyd 1992). Thus when the ferry left port, with its bow doors open on March 6, 1987, water began to flood through the open bow doors across the main car deck, which, as with other ro-ro ferries, lacked dividing bulkheads. The Herald quickly became unstable, and capsized in about 90 seconds. The sea was calm and there was only a light breeze, 193 passengers and crew members died.[10]

The Channel between England and continental Europe is one of the world's busiest waterways. During the 1980s, passenger ferries were facing the threat of new competition. Sealink UK Ltd had been privatised by the neo-liberal Thatcher Government in 1984, opening the market up to greater competition between operators. Reductions in crew numbers, modernisation of port facilities to speed ferry turnaround time, building of new, larger ro-ro ferries, and a range of cut-price fares began to characterise the market. And, after 1986, such trends were exacerbated when British and French Governments agreed to construct the Channel Tunnel, to be operative by 1994.

The findings of the Court of Formal Investigation into the *Herald* capsizing, held in 1987, were remarkable for their content and tone. It is worth quoting one key, and infamous, passage at length:

> ... a full investigation into the circumstances of the disaster leads inexorably to the conclusion that the underlying or cardinal faults lay higher up in the company. The Board of Directors did not appreciate their responsibility for the safe management of their ships. They did not apply their minds to the question: What orders should be given for the safety of our ships?
>
> The directors did not have any proper comprehension of what their duties were. There appears to have been a lack of thought about the way in which the Herald ought to have been organized for the Dover–Zeebrugge run. All concerned in management, from the members of the Board of Directors down to the junior superintendents, were guilty of fault in that all must be regarded as sharing responsibility for the failure of management. From top to bottom the body corporate was infected with the disease of sloppiness .... It

> is only necessary to quote one example of how the standard of management fell short .... It reveals a staggering complacency. (Department of Transport 1987: Para 14.1)

The disaster had significant legal consequences. The inquest held later in 1987 into the capsizing returned verdicts of 'unlawful killing' – despite the coroner instructing such a verdict inadmissible – and created the possibility of a criminal prosecution. A manslaughter case – against P&O European Ferries, owners of the *Herald*, and seven individuals – did follow, but the judge closed proceedings before the prosecution had finished presenting its case on the basis that the legal test under current manslaughter law would not be met in this case. Indeed, whilst avoiding criminal sanction, the chairman of P&O European Ferries, Sir Jeffrey Stirling, became Lord Stirling in Prime Minister Thatcher's 1990 resignation honours.

This case was significant for several reasons. It had a symbolic significance, in several respects: due to the sheer scale of the tragedy, due to the images of it that pervaded the media of this enormous, everyday piece of technology capsized, and also due to the irony of its name – *Herald of Free Enterprise* – so that its sinking could be viewed as emblematic of the problems created for public and worker safety by a Government-led enterprise culture which seemed to valorise profit *and* risk taking. Further, in the failure of the prosecution, the problems of successfully holding large, complex organisations to account for the production of death was raised up political and popular agendas. The ensuing outcry was one factor leading, in 1996, to new proposals for a new law on 'corporate killing' – though at the time of writing, this has still to pass through the UK parliament.

Design changes to ro-ro ferries had been called for in 1985, following the capsizing of the vessel *European Gateway* off Harwich in 1982. Six people were killed in that incident. Following Zeebrugge, the United Nations body, the International Maritime Organisation (IMO) convened a special conference and eventually passed the Safety of Life at Sea Regulations (SOLAS 90). Regulation 8 of SOLAS 90 states that all ro-ro ferries must be able to stand upright long enough for passengers to evacuate. Although this applied automatically to vessels built after April 1990, it did not apply to those built before that date – it was argued that the cost of installing bulkheads on existing craft would, if passed on to passengers, increase fares by up to 35 cents per ticket. The standards agreed as suitable in SOLAS 90 were repeatedly postponed. They came into force in October 2005.

Yet ro-ro ferries seem inherently unsafe. Between 1989 and 1994, Lloyds register recorded 4,583 lives lost at sea – a third of which, 1,544, were lost in incidents involving ro-ro ferries, despite these being but a small fraction of the world's fleet. Then, infamously, on 28 September,

1994, the ro-ro ferry *Estonia* capsized in the Baltic sea killing – officially – 852 people (small children were not registered as passengers so the toll could have been significantly more). Water entered the vessel through defective bow doors. Most of these people died in the vessel because they did not have sufficient time to get on to the upper deck. Six weeks after this sinking, a report from the Department of Transport's Marine Safety Agency discovered that of the 107 ro-ro ferries it had inspected, bow door faults were found in one in three vessels.

In 2002, P&O announced a widespread reorganisation, cutting its fleet of ships from 24 to 17 and ending several 'uneconomic routes' – including services from Dover to Zeebrugge. P&O's ferry services director, Graeme Dunlop, stated at that time, eight years after the opening of the channel tunnel that, 'We knew the tunnel was going to open and we knew it was going to grab a share of the market. Our strategy was to increase the size of the market, so we each got a slice of a bigger pie'.

Piper Alpha, July 1988

Key causal issues highlighted in this case include: casualised/sub-contracted labour; aggressive management; industry norms; systematic failure to heed warnings; political economy of speed; global market pressures; failings in legal and regulatory processes (see Table 1).

Around 10 p.m. on 6 July, 1988, a series of explosions tore through the Occidental-owned Piper Alpha oil production platform in the Northern Sector of the British North Sea. One hundred and sixty seven were killed on Piper Alpha; 61 workers survived. Thirty bodies were never recovered from the North Sea after Occidental declined to pay for the excavation of the platform from the sea bed.

Subsequent accounts of events following the explosions indicate a high degree of management contribution to the events. Negligent emergency provision and under-maintained critical safety systems made a catastrophic situation worse: emergency lighting failed; there were hardly any torches available to the crew; each of the lifeboats was located in the same section of the platform, which also happened to be inaccessible; and no provision had been made for an alternative escape route to the sea. Most people on the platform gathered at the emergency muster point, the accommodation module, which due to its location above the gas compression module also happened to be one of the areas on the platform most exposed to fire and explosion. The accommodation module, constructed from wood and fibreglass, quickly began to burn. The water deluge system – the platform's main defence against fire – failed. When two life rafts were launched, they failed to inflate. The standby safety vessel, a converted fishing boat, had no medical supplies to treat survivors as they were pulled from the sea, and the Tharos, Occidental's state-of-the-art floating fire engine, could not muster sufficient water pressure to reach the flames.

Occidental's senior management had been warned by their own consultants that the platform would not withstand prolonged exposure to high intensity fire. It was a warning that they chose to ignore since the company's assessment of the risk of such an incident occurring did not justify the expense of refitting the platform with high-grade fireproofing. The sequence of events that followed the initial explosions illustrate the strictly observed 'production first' dictum of offshore management. Managements on platforms connected by the same pipeline chain, the Tartan and the Claymore, declined to shutdown production, and instead continued to feed the blaze with oil as they obediently awaited permission for a costly closure of the pipeline from senior management onshore. There was no prosecution of Occidental, a firm that already had a criminal record for killing one of its workers in September 1987.

Piper Alpha, rather that being understood as a one-off, unpredictable event, should be understood as the culmination of a generalised safety crisis in the industry. The origins of this crisis are in the failure of both the regulatory system and the operating oil companies' management regimes. Also of key significance was the political economy that ensured the oil was to function as the motor of the neo-liberal restructuring of the UK economy in the 1970s and 1980s. This had three principle effects. First, as Carson (1982) argued, the conditions within which British oil policy developed are best described as 'the political economy of speed' whereby the speedy development of an industry – central to offsetting Britain's economic decline in the 1970s – came at great cost to the workers, measured in abnormally high rates of injury and death in the sector (11 times the fatality rate in the construction industry and nearly nine times the rate in mining). A related effect was regulatory acceptance, the 'institutionalisation of tolerance' of this appalling toll, indicated by the legal anomalies, jurisdictional gaps, and low level of resources available to the regulatory regime.

Second, the particular features of the labour process in the offshore oil sector were a major contributory factor. The use of casualised and subcontracted labour was quickly established by the offshore drilling and operating companies in the early years of the industry. The legacy of this structure was that, over the years, the percentage of offshore workers employed as sub-contractors has remained at between 80 per cent and 90 per cent. In the pre-Piper Alpha period, workers were usually employed on short-term contracts that often lasted no longer than a few weeks or months, and contained few contractual rights. Trade union organisation was virtually unheard of. Brutal management styles ensured that those who were found to have trade union sympathies, or those who vocally expressed concerns about safety were routinely 'NRBd' (told they were not required back). Eighty three per cent of workers on board Piper Alpha on the night of the disaster were sub-contracted. The marginalisation of workers' expertise and knowledge of safety in the offshore management regime was to have catastrophic consequences. In the months preceding the disaster, management ignored a series of reports of gas leaks by workers on the Piper Alpha. On the day of the disaster, two workers had complained about a gas smell. Although one received permission by the platform safety officer to down tools, management had apparently declined to interfere with work routines and conduct an investigation.

Third, the contractual subordination of workers as a result of the offshore labour market structure had been exacerbated by a chain of events in the market which shook the industry in the mid-1980s. The collapse of the OPEC cartel quota system in 1985 (the average oil price per barrel plummeted from more than $30 in November 1985 to around $10 in April 1986) had a dramatic effect on the industry. In order to defend profit levels, oil companies slashed their operational budgets by between 30 and 40 per cent across the board. The impact on the workforce was devastating. Wage levels fell dramatically and 1986 saw up to 22,000 jobs lost in the industry. The operators' response to the oil price crash had far reaching implications for workplace safety in the industry. Funding allocated to ensuring the regular maintenance of plant equipment suffered the same fate.

Fatality and injury rates remain abnormally high in the industry, and, despite the introduction of some limited improvements to hardware and, a new regulatory regime following the official inquiry into the disaster, the workforce remains largely casualised and lacks trade union safety rights. In the years since Piper Alpha, the workforce has remained under pressure as a result of successive cost-cutting exercises aimed at recuperating the costs resulting from Piper Alpha.

Simon Jones, April 1998

Key causal issues highlighted in this case include: inter-firm relationships; casualised/sub-contracted labour; government sub-contracting out of employment responsibilities; changes in welfare state; failings in legal and regulatory processes (see Table 1).

In April 1998, Simon Jones signed on for casual work in Brighton with a local employment agency, Personnel Selection. According to Emma Ainsley, a friend, Simon had taken the job 'to get the dole of his back'.[11]

Simon was required to register with an employment agency, Personnel Selection, whose job it was to find him work. Under the Job Seekers Allowance scheme – part of New Labour's broader Welfare to Work strategy – claimants must continually demonstrate availability for and willingness to work; these conditions make refusing offers of work liable to lead to a withdrawal of state 'benefits'. Simon's first job with Personnel Selection was at Shoreham Docks, working for Euromin Limited, a Dutch cargo company. He went to work in a ship's hold, unloading its cargo. Within an hour of arriving for his first day of work, he was dead. His head had been crushed and partially severed when a three tonne 'crane' grab closed around it.[12] The grab should not have been there; it certainly should not have been open. The work required chains which should have been fastened to a hook instead. Changing back between a grab and a hook costs time and therefore money. Ten weeks prior to the incident, under Mr Martell's instruction, the chain had been welded to the inside of the grab so that the crane could be used for either purpose without delay (Kelso 2001). After Simon's death, co-workers were immediately required to clear up blood and debris from the bags of aggregates that were being shifted in the hold, and then to continue working. Emma Aynsley has since said that

> Immediately that me and Simon's family found out that he was working inside of a ship we all felt, as his mother put it, this wasn't an accident, because accidents are things that are avoidable and Simon should never, ever have been put in that situation.

As one of the campaigners in the Simon Jones Memorial Campaign said subsequently, in a strikingly clear analogy, it was like asking someone without a driving licence to drive an articulated lorry.

No prosecution was ever taken against Personnel Selection, though they are covered by the legal requirement to ensure the suitability of work which they assign or offer. Initially, the CPS also declined to prosecute Euromin, the firm for which the student was working (at just over £4 per hour). This is despite the fact that on visiting the scene of the death, the HSE issued two Prohibition notices – regarding the use of the crane – and an Improvement notice requiring the training and supervision of new workers.

A protracted and high-profile campaign by the Simon Jones Memorial Campaign developed. This involved a range of methods: lobbying via letters and phone calls; when the CPS and DPP initially declined to proceed with any case against Euromin or James Martell, legal pressure from lawyer Louise Christian – and an eventual judicial review of the decision not to prosecute; and various forms of direct action, such as occupying the Department of Trade and Industry and the offices of Personnel Selection, picketing the CPS offices and closing Southwark Bridge and the road outside the headquarters of the HSE in London.[13] In March 2000, at judicial review, two high court judges overturned the decisions of the DPP and

CPS not to prosecute. They stated that these had both behaved 'irrationally' by concluding that there was no realistic chance of conviction, and that their decision was based upon a misinterpretation of the law; they were directed to reconsider 'with dispatch' (Brooks 2001). In December 2000, the DPP reversed its decision and announced that James Martell and Euromin Limited. were to be tried for manslaughter and corporate manslaughter. On 29 November, 2001, Martell and Euromin were cleared of manslaughter; Euromin was fined £50,000 for two H&S offences. During the trial the judge described Euromin's and Mr Martell's attitude to safety as 'absolutely deplorable' and as giving 'wholly insufficient thought and attention' to safety. He also stated that the method of attaching bags to a hook welded to the grab should never have been used (ibid.).

Despite this, in his summing up the judge emphasised Martell's good character and, typically in cases of business offending where prosecutions are rare so that there is a high chance that companies and individuals will have no previous convictions, his 'clean record'. Speaking outside the Old Bailey after the verdict, Tim Jones, Simon's brother, criticised the judge for accepting Martell's claims of good character:

> Martell who, from the witness stand, said he would willingly send one of his own sons or his own daughter to work in the same conditions that killed my brother – is that his idea of a good character? He said that because Martell had no previous convictions, that he was unlikely to offend again, and that because he had no previous convictions, that made him a good character. (see also Brooks 2001)

In many ways, Simon's death was a routine killing. Hundreds of workers and members of the public die each year in Britain in work-related incidents. Rarely do these deaths make the headlines – unless they come in the form of 'disaster', that is, a multiple-fatality incident. But if it was routine, then Simon's death was also exceptional in several ways, certainly in its aftermath in terms of the campaign that was fought by the Memorial Campaign (friends, family, safety campaigners); and in successfully forcing the DPP and CPS to about-turn on their refusal to prosecute, it was far from routine. And, even though the case 'failed', a prosecution for manslaughter following a death at work is rare indeed – at the time of writing there have been only about 40 cases taken against individuals and/or companies in English and Welsh legal history;[14] most have been unsuccessful (see www.corporateaccountability.org/manslaughter/cases/main.htm).

The Simon Jones Memorial Campaign based their fight around the issue of casualisation – a growing feature of working life in neo-liberal Britain. For the campaign, casualisation translates into people being 'forced into low paid jobs with little or no training, no job security, no sick pay and no holiday pay means bigger profits for companies'. As one of the campaigners stated, 'Ten years ago, if you were going to work in a dock, you would had to have some training and knowledge. Nowadays, with the growth of casualisation, the search for cheap and throwaway labour, you have people who are doing work on docks that they have no training to do'. Casualisation, short-term employment, and agency work are all common features of the deregulated labour market. They are also furthered by a benefits system which forces claimants to take work – even work for which they are patently 'unfit' – on threat of withdrawal of financial support from the state. Finally, the role of Personnel Selection – acting as the 'middle-man' between the state and Euromin – is also symptomatic of a state contracting out its functions to the private sector. In short, Simon's death is only explicable in the context of neo-liberalism. It was quite literally neo-liberalism that, to refer back to Emma Aynsley's words, above, 'put [Simon] in that situation'; prior to neo-liberalism Simon Jones simply could not have been where he was to lose his life: in the heyday of the welfare state, there would have been no compulsion to work in exchange for benefit entitlement, no role for private companies in finding that work, and no chance of him working on the docks without having been certified as competent to do so under the national Dock Labour Scheme (Lavalette and Kennedy 1996). In other words, although a routine killing, Simon's death is only comprehensible in the context of wider social, political and economic trends. The vulnerability which exposed Simon Jones to circumstances that led to his death was created through structural factors.

Transco, December 1999

Key causal issues highlighted in this case include: poor design, maintenance, and management; market pressures; systematic failure to heed warnings; victim blaming; failings in legal and regulatory processes (see Table 1).

On the 22 December, 1999, Andrew and Janette Findlay and their children Stacey and Daryll were killed when a gas explosion completely destroyed their home at 42 Carlisle Road in Larkhall, Scotland. When investigators turned up at the wreckage of the house, they closed the 'slam shut' system to close off the supply of gas into the site of the explosion, but gas fires continued to burn as a result of 'gas coming up from the ground' (*BBC News*, 18 February, 2005). Readings taken at the front of the house registered 100 per cent gas. The explosion was conclusively found to have been caused by a leaking iron gas main which was subsequently found by investigators to have been severely corroded and had a total of 19 leaks (ibid.). Gas leaking from the mains had found its way under the kitchen of the house and ignited.

The prosecution in a subsequent court case alleged that Transco had failed to properly monitor its own records which showed that the mains had leaked 27 times. It was also recorded that escapes of gas at Carlisle Road had been reported on 13 occasions between July 1988 and December 1999. In a subsequent case on the 22 October, 2000, a couple in Dundee were killed in a gas explosion caused by a fractured iron gas main, the operation and maintenance for which Transco was responsible. Despite the close similarity with the Larkhall explosion, no criminal charges were laid against the company. Instead, the company came to an agreement with the HSE the following year on a new timescale for replacing cast iron mains piping (Health and Safety Executive 2004a).

Faulty gas mains are likely to continue to cause similar incidents in the UK. Most iron gas mains date from the Victorian and pre-war eras. The previously nationalised British Gas began to replace the corroding iron mains in the 1970s and since privatisation, this has primarily been the responsibility of Transco, the single largest gas supply company. In the 1980s, Transco began replacing the pipeline system with iron ductile mains which were subsequently also found to be at high risk of corrosion.

Britain's corroding system of gas mains have not been helped by Transco's failure to both replace and maintain its pipeline network. In the year following the Larkhall explosion, Transco were criticised by the industry regulator Ofgem, for failing to reach its target to replace corroding pipelines by 29 per cent (Hodge 2002). In 1997, Transco laid off 1,000 maintenance workers. Four years later, it admitted that it needed approximately 1,000 more workers to maintain the safety of the gas supply network (Lawrence 2001). In early 2002, Transco announced a 17 per cent reduction in its staff, cuts that amounted to the loss of 2,400 jobs.

In February 2002, Transco became the first company in Scotland to be charged with culpable homicide, the Scottish equivalent of the offence of corporate manslaughter in England and Wales. The prosecution had proceeded on the basis of a High Court ruling that it was not a requirement in Scots law for an individual to first be identified before the company could be found guilty of culpable homicide (see the *Herald of Free Enterprise* case). However, a Court of Appeal ruling dismissed the Crown's case that corporate liability could be imputed by aggregating several decisions that took place in the company in several different meetings by different company officers. The Court of Appeal effectively ruled that a corporate homicide case in Scotland also rests on the principle of identification and dismissed the charge of culpable homicide against Transco (see www.corporateaccountability.org/manslaughter-scot.htm). The failure of the corporate homicide case, the first of its kind in Scotland, appears to have severely dented the prospect of subsequent cases of corporate homicide being prosecuted, even in cases of small or medium-sized firms where identification would be less of an obstacle. There have been no attempts to prosecute such a case since. In December 2004, the Scottish Executive announced their intention to introduce a new offence of corporate killing.

Although the homicide case failed, the thoroughness of the corporate homicide investigation between the HSE and the police revealed serious failings in the company's management of risk to the public. This was important in a the subsequent prosecution of Transco for breaching the Health and Safety at Work Act at Larkhall. In this case, the company attempted to attribute blame for the explosion not to their own failings, but to a leak inside the house. It was an attempt to deflect responsibility that was not looked upon favourably by the court. Sentencing Transco in the High Court, Lord Calloway noted:

> the company have chosen to attempt to blame the explosion on an internal pipe leak (that is, something for which they are not responsible) despite overwhelming evidence to the contrary including the views of their own employees on the site after the explosion. There was no evidence at all in this case that such an internal leak had occurred. That aspect of the defence by the company serves only to demonstrate that the corporate mind of Transco has little or no remorse for this tragedy which, they ought at least now to accept, was exclusively of their own creation (*BBC News*, 25 August, 2005).

The company's records falsely showed that the mains outside the Findlay's house in Larkhall had been replaced by plastic pipes (*BBC Frontline Scotland*, 29 November, 2000).

The company was fined £15 million, a record fine for a health and safety offence, which Calloway described as '... an appropriate penalty for a serious failure over many years of one of our privatised utilities, a company in whom the public put their faith and to whom they pay substantial sums of money to ensure the safe transportation of their fuel'. The fine, despite its size, equated to less than 2 per cent of the company's post-tax profits for the previous year. In any case, because the company is able to choose how to spread the cost of the fine, the burden could be passed on to its consumers or workers in the form of higher charges or job cuts. Following the imposition of the fine, a company spokesperson declined to apologise for its crime (The *Observer*, 28 August, 2005).

Morecambe Bay Cockle Pickers, February 2004

Key causal issues highlighted in this case include: power and reach of multinational corporations; casualised/sub-contracted labour; legal/illegal business relationships; systematic failure to heed warnings; victim blaming; global market pressures; changes in welfare state; failings in legal and regulatory processes (see Table 1).

On 5 February 2004, 23 cockle pickers were drowned at Morecambe Bay on the Lancashire coast. This was the worst single industrial disaster in Britain since Piper Alpha. The dead were all immigrant workers from China who had been put to work on the highly profitable cockle beds of the Fylde coast. Two of the bodies were never found.

The tides in Morecambe bay are notoriously quick and the area was known for difficult currents and quicksands. On the day of the disaster, the tide times allowed only three hours of work before dusk. Because of high tides and weather forecasts, most of the British groups of cocklers had left the Bay at 5 p.m., the time that the Chinese workers arrived. The Chinese cocklers were organised by a 'gangmaster', Lin Liang Ren, who was responsible for feeding and housing the cocklers and supplying them with waterproof clothes and tools. On the night of the disaster, Lin Liang Ren drove the cocklers across the mud flats, several miles from the shore. The cocklers were caught in the dark in the midst of rapidly rising tides and although they managed to use mobile phones to call the shore, only one worker was rescued from the water. In March 2006, Lin Liang Ren was convicted for the manslaughter of 21 cockle pickers. The Liverpool Bay Fishing company that bought the cockles from Lin to sell on to larger producers was cleared of facilitating the crime.

There had been several clear warnings that had gone unheeded. In June 2003, the local MP Geraldine Smith had written to the Home Office with concerns that inexperienced Chinese cocklers were being employed on a fifth of the wages of British workers and that they were being exposed to dangers that experienced cocklers would not allow. A near miss, when 40 workers had to be rescued just six weeks before the disaster was reported and widely known about. Although the dangers to which Chinese cocklers were being exposed had been brought to the attention of the industry association responsible for issuing permits (the North Wales and North West Sea Fisheries Committee), no action was taken by the industry to limit those risks (*Liverpool Daily Post*, 7 December, 2005).

An ongoing dispute between British and Chinese gangs of cocklers intensified during the months and weeks leading up to the disaster. The dispute was partly about the territorial rights to work the bay, but was also fuelled by the assumption that Chinese labourers were more likely to take risks than locally-established cocklers. In the week before the disaster, buyers were being asked by British cocklers not to purchase cockles harvested by Chinese gangs (Herbert and Nash 2004).

The key legislative response to the Morecambe Bay disaster was the Gangmasters (Licencing) Act which passed into law in July 2004. The Act was supported by the government, the trade unions, all of the major supermarkets and the National Farmers Union. The new law stipulated that 'gangmasters' (employers that organise gangs of casual labourers, normally paid on a piece-work basis, in the agricultural industry) must now formally register with the Gangmasters Licensing Authority before they can operate. The Act created a skeleton regulatory structure for gang labour, but it did not specify a legal minimum for safety conditions, training or housing for immigrant workers. In the UK regulatory system, the HSE is responsible for monitoring safety standards in both illegal and legal enterprises. In practice, however, HSE has neither the resources nor the political will to inspect and investigate unregistered enterprises and seek compliance or take enforcement action in this sector. In the immediate aftermath of Morecambe Bay, the HSE carried out inspections and drew up safety guidance for the cockling industry. Yet the regulatory agency lacks the capacity to provide a sustained regulatory presence in the agricultural industries in which immigrant workers are concentrated.

The Trades Union Congress (TUC) estimates that around 2.6 million immigrant workers currently work in the UK and that only 1 per cent of newly immigrant workers are members of trade unions (Trades Union Congress 2003a). Lack of trade union membership is compounded by a lack of rights to minimal financial support. Immigrant workers are therefore easily recruited into the most casualised, low paid and dangerous work. The Transport and General Workers' Union estimates that after accommodation and travel, some migrant workers earn as little as 78p per hour and it is normal for wages to be paid below the legal minimum (*BBC News*, 25 February, 2004). It is likely that most of those that died had to been forced to work in the industry in order to repay the expenses of their passage to the UK. According to evidence revealed in court all of those who died had been smuggled into Britain and some owed up to £20,000 to people-smugglers (The *Guardian*, 25 March, 2005). In this sense, their position as workers can be described as 'indentured labour'. The immediate response of the UK government was to blame the workers for coming here to work in the first place. Home Secretary Charles Clarke and other government ministers argued that an ID card system would have protected migrant workers like those who were killed at Morecambe Bay because it would have prevented them entering the country and remaining here undetected (*The Times*, 20 December, 2004). Clarke's response echoed that of the tabloid newspapers in Britain which explained the cause of the disaster as the rise in illegal immigration as opposed to the absence of any safety controls. Morecambe Bay was therefore read as an 'immigration' or 'asylum' problem, rather than a failure of the health and safety regime.

A dramatic change in market conditions in the global market for cockles also helps us understand the broader economic pressures which speeded up the harvest. In 2002, an oil tanker, the *Prestige*, sunk off the west coast of Spain resulting in a massive oil slick. There was a subsequent banning of all fishing and shellfish harvest along the western and northern coasts of Spain which created a rise in the demand for cockles in the Spanish market. The Spanish company Conservas Dani bought a family-run firm based in Wales, Jones, in 2002 which had previously preferred to deal with established cocklers. Conservas Dani's arrival in Britain intensified demand and created new opportunities for gangmasters employing Chinese workers (Herbert and Nash 2004). The Morecambe Bay disaster can therefore be understood as resulting partly from pressures linked to changes in global markets and the role of transnational corporations in intensifying those markets.

The risks to workers in Morecambe Bay remain acute. In March 2006, more than two years after the disaster, the Royal National Lifeboat Institute reported that half of its 100 emergency call-outs in the Morecambe Bay area since the 2004 disaster were cockling related (*BBC News*, 24th March, 2006).

Sonae

Key causal issues highlighted in this case include: state support for investment of company; frequent offending ('recidivism'); safety and environmental offences; failings in legal and regulatory process; aggressive management; perceived conflict between safety and jobs (See Table 1).

In 1999, the Duke of Edinburgh opened a chipboard factory, run by Sonae UK, in Kirkby, Merseyside. As well as being the world's largest manufacturer of wood panels, Sonae – a Portuguese-based conglomerate – has major commercial interests in retail, telecommunications and a range of services. The new factory was located in Northwood, a poor ward in a poor town (Kirkby) in a poor borough (Knowsley). For example, in the Office of Deputy Prime Minister's list of the worst wards in the UK for employment in 2000, three wards in Knowsley – Longview, Princess and Northwood – were listed in the worst six, while on the Index of Multiple Deprivation, which includes indices for crime, health, housing, education and access to local services, Longview was fourth, Princess eighth and Northwood twentieth. Little wonder that the plant – employing 200-plus workers – was established with the financial and political backing of several public authorities, including Knowsley Borough Council, nor perhaps that it was awarded a grant of £1.95 million, by Peter Mandelson, as Secretary of State for Trade and Industry. The new factory had enormous economic, social and political significance for a regenerating town and region.

However, since its opening, the plant has been beset by health concerns among workers and by environmental fears among residents, and indeed a series of legal actions against it. For example, in 2003, the company received a fine of £37,000 after pleading guilty to five pollution offences.[15] Alongside these problems, the plant also has an extremely poor safety record – upon which this case study shall focus. A listing of key incidents and enforcement action illustrates this clearly, while also making it hard to resist the conclusion that, if the *Oxford English Dictionary* defines a recidivist as 'one who habitually relapses into crime', then that label applies to Sonae – even if it is almost entirely reserved, in popular and academic usage, to individual, lower-class offenders.

As early as April 2000, the HSE had instituted a prosecution after a Sonae employee, Ian Fairclough, had become trapped in the clamping mechanism of a Dieffenbacher hydraulic press. He suffered serious crush injuries to his arm and chest when he entered the line to release a trapped board and became trapped himself when the line activating the board was released. Sonae was fined £15,000 and ordered to pay £16,703 costs (*Liverpool Echo*, 25 February, 2003). Mr Fairclough was off work for more than three months, was left with a disability down his right side and – three years later, at the time of the prosecution – still needed treatment for post traumatic stress disorder. The judge said: 'It is important that firms such as Sonae do not sacrifice safety for profits. In this case, more care should have been taken and more precautions put in place to ensure such accidents do not happen' (ibid.).

Between October 2000 and April 2001, six over-three-day injuries and seven major injuries[16] at the site were reported to the HSE. In April 2001, Michael McNamara broke a leg after he caught it in a piece of machinery he was trying to unblock. On 20 May 2003, Sonae was fined £35,000 and ordered to pay £6,417.90 in costs, following successful prosecution by the HSE for two offences under the Health and Safety at Work Act 1974 related to this incident. Ian Connor from the HSE said:

> It is a lesson not just for Sonae, but for all employers to assess the risks and make sure proper safeguards are in place, as well as making sure employees are adequately trained and supervised ... The combination of a lack of safeguards with the lack of training and supervision left this an accident waiting to happen.[17]

In September 2001, there was a large fire at the factory, to which seventy firefighters attended. In the early hours (around 5.15 a.m.) of 17 December 2001, there was an explosion followed by a fire at the plant. No one was injured (*Liverpool Echo*, 17 December, 2001). Then, in June 2002, there was another dust explosion at the Sonae plant. Worker John Thomas's life was 'saved' when firefighters took him from the fire, though he was taken to hospital 'seriously ill' (*Liverpool Echo*, 3 June, 2002) with 'head, chest and back injuries' (*Liverpool Daily Post*, 27 July, 2002). He subsequently recovered. On 10 July 2006, Sonae was fined £70,000 and ordered to pay £77,046 costs at Liverpool Crown Court after pleading guilty to a charge arising from this incident. At the conclusion of the case, the HSE noted that as well as Mr Thomas being injured, 'many others were placed at risk and very substantial damage was caused to the premises'. HSE Inspector Tim Beaumont, who headed up the investigation, added, 'The basic problem uncovered by our investigation was that during the design and construction of the factory in 2000, Sonae did not take an overall view of safety in connection with the manufacturing process.'[18]

Safety 'problems' continued after the two fires and explosions in 2001. On 14 June 2002, a worker was run over at the plant by a reversing forklift truck. This led to a HSE prosecution in December 2004, which resulted in a fine of £12,000 and costs of £13,099.[19] In addition to the incidents noted above and the four HSE prosecutions, the HSE has also taken formal enforcement action at a quite staggering rate given what is known about the preferred modus operandi of the HSE and its inspectorates (see Chapters 4 and 7):[20]

- in May 2001, an improvement notice was issued on work in confined spaces;
- in September 2001, a prohibition notice was issued on the wood chip fuel feed area following an explosion;
- in November 2001, an improvement notice concerning automatic self-propelled transfer carriages was issued;
- a prohibition notice was issued on 25 June 2002, prohibiting the restarting of the plant involved in a recent explosion until a suitable and sufficient assessment of the risks associated with this plant was carried out;
- also in June 2002, an improvement notice on the safe use of industrial lift trucks was issued.[21]

A further five prohibition notices were served on the plant between 11 July 2002 and 15 October 2002.[22]

Then, in October 2003, a worker became trapped for around an hour, 50 feet in the air after being caught in a conveyor belt (*Liverpool Echo*, 28 October, 2003). He suffered serious leg injuries (*Liverpool Echo*, 29 October, 2003). By October 2003, Knowsley Council were reported as giving Sonae

'42 days to improve or face a £50,000 fine' – and the following summer local MP George Howarth demanded the closure of the factory (*Liverpool Echo*, 30 July, 2004). The factory continues to operate.

Safety crimes in outline

The above cases have been presented schematically only. But in presenting them, not only have we been able to mark out the broad territory with which this text is concerned, but we have also begun to highlight some key, recurrent themes which seem to arise in any *post hoc* analysis of a specific safety crime. These are presented in outline form in Table 1.

Table 1 Key causal issues highlighted in case studies

Case	Key Issues Highlighted
UK Construction	casualised, sub-contracted and increasingly migrant workforce long and complex supply chains legal/illegal business relationships aggressive management market pressures industry norms problems in regulatory process
Bhopal	power and reach of multinational corporations 'developed' versus 'developing' economy safety standards poor design, maintenance, management global market pressures victim blaming power to circumvent/prevent legal process safety *and* environmental offences long-term victimisation – lack of compensation
Herald of Free Enterprise	market pressures industry norms systematic failure to heed warnings poor design and maintenance, management victim blaming failings in legal and regulatory processes

Table 1 continued

Case	Key Issues Highlighted
Piper Alpha	casualised/sub-contracted labour aggressive management industry norms systematic failure to heed warnings political economy of speed global market pressures failings in legal and regulatory processes
Simon Jones	inter-firm relationships casualised/sub-contracted labour government sub-contracting out of employment responsibilities changes in welfare state failings in legal and regulatory processes
Transco	poor design and maintenance, management market pressures systematic failure to heed warnings victim blaming failings in legal and regulatory processes
Morecambe Bay	power and reach of multinational corporations casualised/sub-contracted labour legal/illegal business relationships systematic failure to heed warnings victim blaming global market pressures changes in welfare state failings in legal and regulatory processes
Sonae	state support for investment of company frequent offending recidivism safety *and* environmental offences failings in legal and regulatory processes aggressive management perceived conflict between safety and jobs

We wish finally to emphasise that despite these cases spanning more than 20 years, none of them remain a matter of history. The construction industry, at least in the UK, is booming, and continues to kill and injure disproportionate numbers. Many of the victims of Bhopal continue to suffer in terms of a polluted environment, inability to work, lack of compensation, and inadequate access to medicines. The post-Piper Alpha regulatory regime remains lax, a cost-cutting, bullying environment remains the norm offshore, and injuries and deaths continue. As we write,[23] the story of the sinking of another ro-ro ferry unfolds, this time in the Red Sea, off Egypt, with some 1,000 people dead – and many 'ro-ro' ferries continue to sail the seas. The Simon Jones Memorial campaign continues to fight against casualisation and for effective corporate manslaughter legislation, the latter being a fight that had really begun following the sinking of the *Herald of Free Enterprise*, and an issue on which successive Conservative then Labour Governments have 'committed' themselves, but have yet to introduce.[24] Migrant workers in the UK continue to be exposed to risks on a daily basis. Transco has been implicated in similar fatalities following the Larkhall explosion. The Sonae plant continues to operate, and to attract criticism for standards of safety *and* environmental management. None of these incidents is history in the sense that they have left thousands of grieving relatives.

The themes set out in figure 1will recur throughout the following chapters – and we shall return to them in detail in Chapter 9, our concluding chapter. Of course, these common themes are refracted differently in specific case studies, have more or less significance and combine with other context-specific factors. But taken together, these cases indicate causes that reside in systems of management, in features of wider industry and market contexts which provide the conditions from which those crimes are produced, and in systems of law and regulation, which create opportunity structures for such offences. This is not to deny the role of contingent factors. But overwhelmingly the cases demonstrate – as will be explored empirically and theoretically through the chapters that follow – that to represent these, or safety crimes in general, largely in terms of 'accidents', or by attributing a key causal role to individualised 'human error', simply obscures rather than facilitates understanding, even if this mystification has particular benefits for certain social groups. If, then, this text is about rendering visible that which is obscured, as we noted above, then it is also about engaging directly and critically with the mystifications that separate off the phenomena with which we are concerned from crime, law and order agendas.

To these ends, then, the structure of the book is as follows. Having indicated what safety crimes look like in the above cases, we turn, in Chapter 2, to map occupational deaths and injuries, using various data to indicate the scale and distribution of these phenomena, of which safety crimes are a subset. Chapter 3 then assesses the ubiquitousness, power and significance of a series of related arguments which cast these deaths and injuries so mapped as *accidents* – that is, as quite distinct from crimes. In this chapter we also raise key questions about the failure of criminology as an academic discipline to incorporate safety crimes into mainstream understandings of violence. Chapter 4 then considers the extent to which we can estimate the scale of offending from this pool of death and injury data. While definitive answers are impossible to reach, we conclude that safety crimes represent at least as great a crime problem as those crimes of violence upon which the Home Office typically focuses, namely crimes of inter-personal violence recorded through official crime data.

Chapters 5–8 turn to focus upon the legal, regulatory and criminal justice approaches and responses to safety crimes. Specifically, Chapter 5 sets out, largely through an historical focus, the development of criminal law in ways that re-positioned safety crimes as distinct from 'real' crimes; Chapter 6 then examines the possibilities for re-assimilating safety crimes into existing structures of criminal law. How existing safety law is enforced empirically, and how we can understand such modes of enforcement theoretically, is the subject of Chapter 7, and in considering regulatory enforcement theoretically we are also attempting to grasp the prospects for, and limits of, changes in such practices. Similarly, Chapter 8 considers the empirical and theoretical dimensions of punishing and sanctioning corporations and employers following successful prosecution for safety crimes.

Finally, our concluding chapter sets out to theorise the nature of safety crimes, both through using the fragments of criminological theorising that can be turned to such a task, and by then developing a framework beyond criminology through which safety crimes can be better understood. This schematic political economy of safety crimes seeks also to set out the parameters of any more effective control. For that, unashamedly, is the aim of this book: to render safety crimes more visible not simply to know them better, but to prevent them and respond to them in ways that minimise and mitigate the exacerbation of social injustice that this book demonstrates is their central effect.

Notes

1 See Tombs and Whyte, 2006.

2 Note that although safety is often used in the couplet 'health and safety', the focus of this book is upon crimes of occupational safety *not* health. Crimes against occupational health do, of course, share some characteristics with safety crimes – yet are sufficiently distinctive to warrent separate treatment.

3 Throughout the rest of this book, the acronym HSE will be used instead of 'Health and Safety Executive' except in references to publications, where the full name will be used.

4 Southern and Easter Regional Traders Union Congress (2005).

5 Throughout the rest of this book, the acronym HSC will be used instead of 'Health and Safety Commission' except in references to publications, where the full name will be used.

6 It also has an 'especially poor' (Health and Safety Commission, 2005a: 13) record on ill-health.
7 The figures in brackets indicate accident rates per 100,000 for the GB as a whole 2003/04 (Southern and Easter Regional Trades Union Congress, 2005).
8 These compare to all industry averages which saw a 25% fall in all enforcement notices issued by HSE in the same years, a 26% decline in prosecutions instituted and a 24% decline in convictions secured www.hse.gov.uk/statistics/enforce/index.htm.
9 Case complied from the following sources: Amnesty International 2004; Cassells 1993; Everest 1985; Pearce and Tombs 1998, Shrivastava 1992.
10 Bergman 2000; Boyd 1992; Crainer 1993; Department of Transport 1987; Wells 1993.
11 All unattributed quotes, and much of the material upon which this case is based, is taken from the campaign's website; see www.simonjones.org.uk. Thanks to the campaign for permission to use this material extensively, and to Simon's brother, Tim, and mother, Anne for their help in checking this material.
12 The 'crane' was in fact an excavator which had been modified for use as a crane by welding hooks onto the grab.
13 Many of these actions, and the work of the campaign in general, is documented in the film. *Not This Time – The Story of the Simon Jones Memorial Campaign*; see www.simonjones.org.uk/newvideo.htm for details.
14 There has just been one analogous case under Scottish law, a charge of culpable homicide against Transco following the deaths of four people on December 22, 1999 when an explosion destroyed a family house in Larkhall. The case was dismissed by the trial judge in 2003.
15 'Sonae UK fined £37,500 for polluting brook' (6 October 2003), at http://www.letsrecycle.com/materials/wood/news.jsp?story=2667.
16 Definitions of these categories of injury are given in Chapter 2.
17 http://www.abertaytraining.co.uk/HomePage/healthnews.asp.
18 'Fourth HSE fine for Liverpool company after explosion', *Government News Network*, Press Release, GNN ref 135406P – see http://www.gnn.gov.uk/Content/Detail.asp?ReleaseID=213291&NewsAreaID=2.
19 Source: HSE prosecutions database – see http://www.hse-databases.co.uk/prosecutions/case/case_details.asp?SF=CN&SV=F170000429.
20 See http://ww.hse-databases.co.uk/notices/notices/notice_list.asp?rdoNType=&NT=&SN=F&EO=LIKE&SF=RN&SV=Sonae&ST=N&x=15&y=14, and also see Chapter 4 on the differences between the three types of notice issued by HSE inspectorates.
21 *Hansard*. House of Commons Written Answers – see http://www.publications.parliament.uk/cgi-bin/newhtml_hl?DB=semukparl&STEMMER=en&WORDS=sona%20notic&ALL=sonae&ANY=notice&PHRASE=&CATEGORIES=&SIMPLE=&SPEAKER=&COLOUR=red&STYLE=s&ANCHOR=20710w19.html_wqn6&URL=/pa/cm200102/cmhansrd/vo020710/text/20710w19.htm#20710w19.html_wqn6.
22 Source: http://www.hse.gov.uk/notices/notices/notice_list.asp?rdoNType=&NT=&SN= F&EO=LIKE&SF=RN&SV=Sonae&ST=N&x=19&y=12.
23 The Ferry sank on 3 February, 2006.
24 The Corporate Manslaughter and Homicide Bill 2006 was passing through the parliamentary process as this book was in its production stage (March 2007), but was subject to intense controversy, and threats to pull it by the Home Secretary. See http://www.corporateaccountability.org/manslaughter/reformprops/main.htm

Chapter 2

Mapping occupational death and injury

Introduction

In contemporary societies, work routinely kills workers and members of the public through acute injury and chronic illness. The scale of this routine killing – deaths occur across all industries, all types of companies – is almost incomprehensible. Although the focus of this text is upon safety, so that we are concerned here with deaths from acute incidents, a useful context for our considerations in this chapter is the International Labour Organisation (ILO) estimate that, at a bare minimum, 2.2 million workers die each year through work-related 'accidents' and diseases (International Labour Organisation 2005: 1). This means that annually there are more than 5,000 work-related deaths every day, while for every fatality there are another 500–2,000 injuries, depending on the type of job.

Of the 345,000 workers estimated by the ILO to have died in incidents (as opposed to occupational diseases and exposures) at the workplace in 2002, by far the greatest number, almost 220,000, were in Asia, with China having by far the highest number of deaths of any one Asian country – 73,595 – with the second highest number being India (48,176).[1] In Europe, Russia is the state with the highest absolute number of occupational fatalities – almost 7,000 – followed by Turkey, Ukraine, Poland and Romania. For Woolfson, flexibilisation – or, as he puts it, the creation of 'a deregulated low-cost, low-wage economy, where labour (preferably "union-free") is comprehensively subordinated to the needs of capital' – is the key to understanding why the states of the former Soviet Union mostly have safety records poorer than the rest of Europe: 'figures show that, as a whole, Central and East Europeans are three times as likely to die at work than those in the EU-15 (9.6 per 100,000 persons in employment compared to 3.4)' (Woolfson 2005).

Britain is not, then, one of the most dangerous places to work in Europe. Nor, however, is it one of the very safest, a claim that could be sustained precisely because of our lack of knowledge of the actual level of injuries *within* nations and the lack of utilisable data for comparison *between* nations (see Eurostat 2001; European Commission 2002). More recently, HSE has undertaken international comparisons of fatality and injury data across EU member states and the US for the year 2000, and was able to conclude that while Britain was one of the safest places, it was being caught up by other nation states (Health and Safety Executive 2000a). The fact of Britain's safety record deteriorating in relative terms is borne out by a recent EU survey which concluded that 'Accidents at work are generally decreasing, but not in all Member States'. Thus, while 'Between 1998 and 2001, serious accidents decreased by 6 per cent in the EU and fatal accidents by 21 per cent', this decrease being replicated across some of the newly acceded states such as Poland, Slovakia and Denmark, it was noted that, 'Despite this general downwards trend in most of the Member States, the incidence rate of serious accidents increased in Sweden (+13 per cent, between 1998 and 2001), the United Kingdom (+10 per cent), Spain (+6 per cent) and Ireland (+5 per cent)'.[2] Such comparisons are fraught with difficulty, and comparative analysis of injury data requires a whole series of factors to be taken into account, but such data do rather undermine the HSC and Government's long-term complacency regarding its safety record compared to other European nations.

In this chapter – which seeks to map the scale and distribution of occupational death and injury in Britain – we begin by setting out these officially recorded deaths and injuries, and note at the same time some of the data's limitations. From there, we seek to reconstruct fatality data, to arrive at a more utilisable figure for occupationally-related deaths than that recorded by HSE. The remaining sections of the chapter then consider the distribution of these deaths and injuries, noting their variation in terms of differential vulnerability, and paying particular attention to gender and ethnicity.

Using official injury data

The most common measurement of occupational 'safety' is recorded injury data; data for the United Kingdom are kept by the HSE and are currently collated under the Reporting of Injuries, Disease and Dangerous Occurrences Regulations 1995 (RIDDOR 1995), which place a legal duty upon employers to notify the HSE of injuries (and some 'dangerous occurrences') which are then categorised as 'fatal', 'major',[3] and 'over three-day' injuries, where 'incapacity for normal work [lasts] for more than three days'.

Unfortunately for those who would seek to make use of injury data (and of interest in the context of an employer's legal duty to report injuries) each category of *non-fatal* injury data maintained under RIDDOR is subject to significant under-reporting, a phenomenon highlighted over 30 years ago by the Report of the Robens Committee on Safety and Health at Work (Robens 1972: 135, and Chapter 15, *passim*). Although major injury data are more reliable than data for over three-day injuries, each data set is subject to processes of under-reporting which render them somewhat unreliable as measures of occupational injury (see Nichols 1997). In the late 1980s it was estimated that official data record on average just over 40 per cent of injuries which are notifiable to HSE, and just 10 per cent of injuries to the self-employed (Stevens 1992).

More recently, HSE has begun to make use of the regular Labour Force Survey. The Labour Force Survey (LFS) is a survey of some 60,000 private households in Britain, based upon a systematic random survey design, and carried out by the Office for National Statistics. Since 1992 it has been conducted on a quarterly basis, and HSE has asked four questions on (non-fatal) workplace injuries at each winter-quarter survey since 1993/94; HSE data are available for 1989/90, and then for each year from 1993/94 onwards. LFS data are of interest in their own right, since they indicate a widespread failure on the part of employers to meet the legal duty to report incidents, an offence under safety law. LFS data indicate that 'only one quarter of reportable non-fatal injuries to employees' and 'less than 5 per cent' in the case of self-employed workers, are reported by employers (Health and Safety Commission 1998: 1). These rates vary between 11 per cent in 'Finance and Business' and 17 per cent in 'Hotels and Restaurants' to 62 per cent in 'Public Administration and Defence' to 86 per cent in 'Extraction and Utility Supply'.[4] Whichever estimates are accepted, it is clear that reporting rates are extremely low. It has also been concluded recently that 'small firms are more likely to under-report or not report at all' (Daniels and Marlow 2005: iv). Yet while the LFS is a potentially useful adjunct to, and improvement upon, data collected under RIDDOR in terms of a measure of injuries, it is somewhat unreliable, given that there are just four questions asked at the end of a very long survey on a range of employment issues (Pickvance 2006).[5]

Generally, rates of mandatory reporting which appeared to improve briefly at the end of the 1990s are now in decline again (Daniels and Marlow 2005: 2–3). Moreover, given shifts in levels of non-reporting, these levels of under-estimate render any longitudinal analyses virtually impossible – a task already made difficult by the fact that there have been several changes in reporting requirements since 1974 (see Nichols 1997). Given the recognised 'poverty' of minor and major injury data, the variable social processes out of which they are produced, and the shifting legal categories into which they are organised, these need to be read with a whole series of caveats in mind and thus treated with extreme caution. For our purposes they are useful in telling us *something* about the sheer scale of the problem that we are indicating, the distribution of deaths and injuries by sector and so on, and for making some informed judgements about any trends in their incidence.

We can make several brief observations on the data presented in Tables 2–10. First, and not at all apparent from the data presented in the tables, the majority of the injuries recorded by the HSE are sustained by men rather than women – men are three times more likely to sustain a major injury, and about two-and-a-half times more likely to sustain an over three-day injury (Health and Safety Commission 2001: 56)[6]. To some extent, one might expect this on the basis of persistent gendered occupational segregation, at least horizontally. HSE research has concluded that:

> The higher injury rate in men is partly explained by the occupations, hours of work, and other job characteristics. However, after allowing for these characteristics of jobs, men continue to face a higher risk of reportable injury than women. Men face a 35 per cent higher risk of reportable injuries than women (not explained by the jobs held by men and women). (Health and Safety Commission 2001: 56)

This raises the issue of possible peculiarities in the risks faced by women, and subsequent 'official' attention accorded to these. We return to these issues in a later section.

Second, taking the data in Tables 1–3 together, it is clear that there are large *absolute* numbers of major injuries – around 30,000 such injuries to all workers in any given year – and over three-day injuries, up to and in some years above 130,000 recorded injuries across British workplaces. Workplace injury is far from an uncommon experience. Table 4 indicates that there is also an absolutely high number of injuries sustained by members of the public in any one year, even though movement between years is an erratic one.

Table 2 Reported major injuries[+]

	96/97	97/98	98/99	99/00	00/01	01/02	02/03	03/04	04/05
Employees (number)	27964	29187	28368	28652	27524	28011	28113	30689	30541
Rate per 100,000 employees	127.5	127.6	121.7	116.6	110.2	110.9	111.1	120.4	117.9
Self-employed	1356	815	685	663	630	929	1079	1283	1251
Rate per 100,000 self-employed	38.4	23.3	20.3	19.7	19.2	27.8	32.3	33.9	33.0
All workers	29320	30002	29053	29315	28154	28940	29162	31972	31702
Rate per 100,000 workers	115.1	113.8	108.8	104.9	99.6	101.2	101.9	109.3	107.1

Table 3 Reported over three-day injuries

	96/97	97/98	98/99	99/00	00/01	01/02	02/03	03/04	04/05
Employees (number)	127286	134789	132295	135381	134105	129655	128184	131017	121779
Rate per 100,000 employees	580.1	589.2	567.3	550.9	536.9	513.5	506.5	514.2	471.7
Self-employed	2282	984	849	732	715	917	951	1114	1143
Rate per 100,000 self-employed	64.6	28.1	25.2	21.8	21.8	27.5	28.4	29.5	30.2
All workers	129568	135773	133144	136113	134820	130572	129135	132131	122922
Rate per 100,000 workers	508.7	514.8	498.8	487.3	477.1	456.7	450.7	451.5	415.2

Table 4 Injuries to members of the public

	96/97	97/98	98/99	99/00	00/01	01/02	02/03	03/04	04/05(p)
Non-fatal *	35694	28613	23800	25059	20839	14834	12793	13679	14316

P is Provisional

* Injuries which result in the injured person being taken directly to hospital (Health and Safety Commission 2005a: 21).

\+ Injury data presented in the tables in this chapter is based upon HSE and HSE data, in addition to personal communication with HSE Statistics Unit, 26/10/2006, 20/04/2006. See www.hse.gov.uk/statistics/causacc/tables.htm, Health and Safety Commission 2000, 2003, 2005a, 2006b; Health and Safety Executive 2003b.

Third, Tables 2 and 3 both indicate that the recorded differences in fatality rates between workers and the self-employed (above, and Table 6) clearly do not hold with respect to injuries: Table 2 indicates that the rate of recorded major injuries to employees is some three to six times higher than that for the self-employed. This is clearly a reporting effect – if a self-employed worker *dies* then this is likely to be recorded, whilst an injury, even a major injury, is not, since there is no financial or other incentive to do so. Similar observations are relevant for the disparities in Table 3, and here we would expect the disparities in rates to be even greater given the far lesser seriousness of injuries. The data in Tables 2 and 3, then, need to be read in terms of non-reporting, bearing in mind the observation above, that only between 5 and 10 per cent of injures to the self-employed are actually reported.

Fourth, it again seems that over the time period covered there are no unequivocally clear, discernible trends, certainly given the problems of variable levels of reporting. In terms of those trends, we can see from Table 2 that there has been a small decline across the period in the rate of reported injury to both employees and the self-employed; there are slight decreases in rates across the period for both employees and the self-employed – though again, we would stress that the data for the self-employed are almost non-utilisable given that they are subject to up to 95 per cent under-reporting.

Now, notwithstanding the enormous problems in determining accurate figures for the numbers of occupational injuries, one consistent claim has been that the recording of fatal injuries is at least reliable (Health and Safety Commission 1996: 1, 1997: 1; Nichols 1989: 543, 1994: 104, 1997: 126). Indeed, the data on *fatal* injuries represents *the* most reliable available data on occupational injury, perhaps even having been improved recently with the introduction of new reporting requirements in 1995. However, as we shall see below, these data are far from complete.

Data for the reporting year 1996/97 is the first to be collected under new reporting requirements, RIDDOR 1995. Of interest in the context of the discussion regarding definitions of 'violence' that is to follow in Chapter 3 is that the new regulations had introduced an expanded definition of occupational fatalities, including inter-personal violence against employees and new categories of reportable occupational fatalities among members of the public by introducing the 'vital' test of 'arising out of or in connection with work' (Health and Safety Executive 1996: 14) in determining what injuries should be recorded as occupational. This phrase is intended to cover 'the manner or conduct of an undertaking', 'the plant or substances used for the purposes of the undertaking', and 'the condition of the premises used by the undertaking or of any part of them' (ibid.: 14). This new, more inclusive, definition simply renders reporting requirements consistent with the substance of the HASAW Act 1974, which requires an employer to 'conduct his undertaking, in such a way as to ensure, so far as is reasonably practicable, that persons not in his employment, but who may be affected, are not exposed to risks to their health and safety'. Under this expanded definition, a total of 593 fatal occupational injuries were recorded for 2004/05.[7]

Table 5 provides a broad overview of occupational deaths that have been recorded since these new reporting requirements were introduced. Several observations can be made about the data presented therein.

Table 5 Reported fatal injuries

	96/97	97/98	98/99	99/00	00/01	01/02	02/03	03/04	04/05
Employees (number)	207	212	188	162	213	206	183	168	172
Rate per 100,000 employees	0.9	0.9	0.8	0.7	0.9	0.8	0.7	0.7	0.7
Self-employed	80	62	65	58	79	45	44	68	51
Rate per 100,000 self-employed	2.3	1.8	1.9	1.7	2.4	1.3	1.3	1.8	1.3
All workers	287	274	253	220	292	251	227	236	223
Rate per 100,000 workers	1.1	1.0	0.9	0.8	1.0	0.9	0.8	0.8	0.8
Members of the public†	367	393	369	436	444	393	396	374	370
Total (workers + public)	654	667	622	656	736	644	623	610	593

† HSE typically notes that about two-thirds of fatalities to members of the public are the result of suicide or trespass on the railways. Of course, these, as all HSE data, exclude fatalities from work-related road traffic incidents.

First, when these data are presented, HSE refer not to deaths including those to members of the public, but rather utilises what has been labelled a 'headline figure' (Health and Safety Executive 1997; Tombs 1999a), referring only to totals of workers (employees and the self-employed). Now, it should be noted that deaths to members of the public are consistently high. But HSE's justification for omitting these from the headline figure is presumably based on some claim that these are not 'real' workplace deaths – in fact, a significant number of these deaths are suicides, and HSE typically notes that about two-thirds of fatalities to members of the public are the result of suicide or trespass on the railways. Nevertheless, this is a rather curious sleight of hand by HSE, since these deaths are reported to and recorded by them as occupational fatalities, and they may well constitute failures of employers to meet legal duties.[8] In any case, such *ad hoc* alterations to what constitutes legitimate injury data are not made in the opposite direction – that is, by counting *in* deaths which are clearly occupational, but which are not covered by HSE's remit. The most notable example in this context are the 1,000 or so fatal injuries sustained whilst driving each year – there is now almost universal agreement that these are occupational fatalities, but they are not reportable under RIDDOR, and are never referred to as possible 'additions' by HSE when their headline figures are released (see below).

Second, these totals are historically low figures. Clearly work is far safer in Britain than it was in the era of early industrialisation when the factory inspectorate was established and described, for example, in Marx's *Capital Volume 1*, or Engels's *The Condition of The Working Class*. Further, it is possible to argue that the HASAW Act (1974) gave a particular impetus to this long-term improvement, not least since the Committee of Inquiry that ultimately led to the Act and the changes it introduced had itself been established in the context of some rises in industrial injury rates (Dawson *et al.* 1988: 9). However, it would be wrong to attribute too much to the Act *per se*. As Nichols has argued, a crucial factor to be taken into account in this long-term improvement was the relative strength of labour vis-à-vis capital (Nichols 1997: 122, and *passim*). Further, the long term downward trend seemed to be an instance of a 'secular tendency' for fatality rates to fall across industrialised countries due to advances in medical care, improvements in communication and transport, knowledge regarding safer working practices, and levels of investment in technology (ibid.: 128). More latterly, and more specifically, injury rates have declined partly as a result of changes in the occupational structure in British workplaces; HSE has noted, for example, that construction is probably the last remaining heavy industry of any size in Britain (Health and Safety Commission 2005a: 13). Since the introduction of the HASAW Act in 1974, for example, employment has shifted significantly from manufacturing to services: three million jobs in manufacturing have gone, while the service sector now employs more than three-quarters of the British workforce compared with two-thirds of workers in 1974.

Third, within the data presented in Table 5, the absolute figures indicate small declines in the numbers of fatalities to employees, the self-employed and members of the public over the nine-year period. In none of these categories, however, is there a clear year-on-year decline in the absolute number of fatalities. As also indicated in Table 2, the rate of non-fatal injuries has also seen a small decline in this period.

Fourth, it should be noted that the rate of fatalities for the self-employed is higher, and typically about double, the rate for employees – an issue that is discussed below in terms of the relative vulnerability of the self-employed vis-à-vis employees.[9]

Finally and most important of all, whilst those figures might allow us to make some very general year-on-year comparisons in the period since RIDDOR was introduced, the problem is that their usefulness is limited because they cannot be regarded as representing anything close to the full picture of deaths and injuries at work in Britain. Whilst fatal injury data are clearly more utilisable than other injury data, because of the relatively lower chances that fatalities will not be reported, and while the requirements of RIDDOR 1995 are an improvement, there are considerable anomalies within the construction of those figures that render them grossly incomplete.

Now, if there are problems regarding fatality data, this can at least be reconstructed and, as we shall argue in Chapter 4, it can also be used to say something about the level of safety *crimes*. But any such reformulations would still exclude the vast majority of potential safety crimes, since incidents that result in fatalities represent only a small minority of workplace injuries, recorded or otherwise. We have also noted that there are enormous problems of under-reporting of non-fatal injuries, whilst variability in reporting rates over time makes any longitudinal analysis difficult, if not impossible. Injury data does, however, remain of *some* use for us. First because it again indicates the sheer scale of the phenomenon with which this text is concerned. And, second, because we can say *something* about the relative distribution of injuries. The latter point will be picked up again later in the chapter. For the time being we discuss how the available data can be more usefully constructed.

Reconstructing official data

The gap in accounting is best illustrated by the stark contrasts between the consistently higher injury rates amongst the self-employed in the context of fatal injury which are not repeated when we look at non-fatal injury data, where the contrast with data on fatalities is clear. Official data shows that being self-employed makes a worker about twice as likely to be fatally injured, whilst at the same time, being self-employed makes it four to five times *less* likely that a worker will sustain a major injury (Tables 2 and 5). This is, of course, a reporting effect and can only be explained by understanding the relative difficulty with which fatalities as opposed to injuries can be hidden. In crude terms, it is more difficult to hide a body than it is to hide a broken wrist.

Having noted this distinction, we should be careful not to simply assume that fatality data are infallible. Indeed, we can identify at least three separate problems with HSE-published data of fatalities which render their use problematic, and any claims regarding their 'virtual' completeness simply wrong.

First, the coverage of RIDDOR 1995 continues to exclude significant numbers of occupationally-caused deaths, for which there do exist officially collected data. Some of these omissions are at the very least curious – notably fatalities relating to the supply and use of flammable gas[10] – but the least excusable, and numerically most significant, is the exclusion of road traffic fatalities which involve 'at work' vehicles. RoSPA (1998) estimated that there are 800–1,000 such fatalities per annum (see also Bibbings 1996), HSE and the Department of Transport have estimated up to 20 such fatalities a week, indicating just in excess of 1,000 per annum (Health and Safety Executive/Department for Transport 2003), while a recent study confirmed the number of deaths at the highest figure in that range, 1,000, concluding that 'Britain's roads are the country's most dangerous workplace' (Trades Union Congress 2005b; see also Campbell 2005). Incorporating existent, omitted data[11] increases the total of occupational deaths from the HSE's recorded figure of 593 for 2004/05 to somewhere close to 1,700 deaths.[12]

Second, the data maintained under RIDDOR with respect to occupational fatalities remain internally incoherent. Anomalies arise, in particular, from inconsistent reporting requirements depending upon: where a fatality occurs; the length of time between sustaining the injury and death (and here there is a disjunction between the regulatory requirement and its actual operation); and whether the fatal injury is sustained by an employee, a self-employed worker, or a member of the public. To provide an illustration of a situation bordering on absurdity: in 2005 a farmworker, having completed his shift, was being given a lift on the back of a tractor when he fell, and died – but since his death occurred outside working hours, it did not get recorded as an occupational fatality ('Work Deaths are Off the Record', *Hazards*, 93, 2006: 31). The net effect of such anomalies is to omit significant, but unquantifiable, numbers of relevant fatal injuries (see Tombs 1999a: 356–9).

Third, and finally, RIDDOR reflects an under-recording of fatal injuries due to social processes of under-reporting.[13] Before a death can be registered, a valid certificate giving the cause of death must be completed and signed by a registered medical practitioner who attended the deceased during her/his last illness; if a death is shown to be 'violent or unnatural', the Coroner is required by law to conduct an inquest. Two studies (Start *et al.*, 1993 and 1995) have highlighted the inability of both hospital clinicians and general practitioners to recognise some categories of reportable deaths. The first (Start *et al.* 1993) indicated that individual clinicians at all grades showed a variable appreciation of the different categories of cases which should be reported, being mistaken in up to 60 per cent of individual cases. Indeed, 'Deaths resulting from accidents were often unrecognised' (ibid.: 1039). In a second study involving general practitioners (Start *et al.* 1995), only 3 per cent recognised all those deaths which should be reported for further investigation; deaths from industrial or domestic 'accidents' were recognised as cases requiring referral by fewer than half of all general practitioners. Both studies indicate that certifying doctors tend to consider only the eventual cause of death rather than the sequence of events leading to death (ibid.: 193).

There can be no doubt that the published fatality data are inadequate: they project only a minimal number of occupationally-caused deaths in Britain on an annual basis. However, from the points of critique that we raise above, it is clear that these data are liable to some reconstruction, through resort to other, official measures of occupationally-caused fatalities. It remains clear that inconsistencies within the reporting requirements themselves, along with various social processes of under-reporting, mean that the HSE's 'headline' figure – that which is trumpeted in press releases but is merely a combination of fatal injuries to workers and the self-employed required by law to be reported to and recorded by HSE – is a dramatic under-estimate of the numbers killed in work-related incidents in the UK each year. For 2004/05, the headline figure would be 223 – the figure for 'All Workers' (see Table 5). The actual figure is somewhere between 1,600 and 1,700. This more accurate figure would include: deaths to members of the public (recorded under RIDDOR), totalling 370 in 2004/05 (see Table 5); deaths incurred whilst driving and 'at work', about 1,000 per annum; and a small number of miscellaneous categories, such as fatalities relating to the supply and use of flammable gas (23 in 2005), and those arising in the course of sea fishing (10 fatalities in 2004) and merchant vesseling (20 fatalities in 2004) in British waters or on British vessels at sea (Marine Accident Investigation Branch 2005: 27, 19). In other words, to obtain a more accurate figure of officially recorded occupational fatalities, we need to apply a multiplier of between seven or eight to the headline figure.

Having attempted to reconstruct the data more usefully, we would argue that a figure between 1,600–1,700 deaths is workable as a *minimum* estimate of annual deaths from sudden injuries. There are, however, two more major caveats to be made. The first is that this figure, as we remind readers throughout the book, does not include the deaths caused by industrial illnesses and exposures. As we indicated in the introduction, the total figure for deaths caused by work in the UK is likely to be in the tens of thousands. Our 1,600–1,700 therefore represents a minimal figure of a small subset of the total death toll. The second is one that all social scientists will be familiar with, namely that quantitative data, let alone 'official' statistics, alone can allow us to make only very rudimentary observations about any social phenomenon (see Levitas 1996). Statistics may allow us to make some comparisons across time and may allow us to make some estimates about scale and incidence, but they can tell us very little about more complex social features of deaths and injuries at work. They can tell us little about the processes by which deaths and injuries are socially distributed or about the processes that produce deaths and injuries in the first place. Nor can they tell us anything about the differential social *impacts* of deaths and injuries amongst different sectors of the population. What the preceding data fail to convey is the fact that risks faced at work are neither evenly nor randomly distributed: not only are such deaths, injuries and illnesses unequally distributed, but this unequal distribution mirrors well-known structural inequalities within nations – between regions, social classes and communities – as well as between nations.

Dimensions of vulnerability

There are good theoretical reasons and some empirical evidence to indicate that these deaths, injuries and illnesses disproportionately fall upon members of the lowest socio-economic groups. If we ask *who* gets killed and injured, the answer is to be found in certain manual occupations – and is comprehensible in traditional class terms (see Table 10). Thus, for example, HSE data indicate that 'process, plant and machine operatives' are 15 times more likely to incur a reportable injury than 'managers and senior officials', and more than ten times more likely than those in 'professional occupations'.

Historically, then, it is perhaps unsurprising that the key source of opposition to workplace risks has been organised labour, and a key political tactic has been to seek more effective regulation, at best regulation involving workers' representatives themselves – in other words, struggles to secure safer and healthier workplaces have historically been classic struggles on the part of labour and its allies to attenuate management's right to manage and expropriate surplus value. This is a key explanatory factor in the long-term historical trend towards safer workplaces within industrialised economies. It also helps to explain why there is no ineluctable trend towards safer working, somehow associated with industrialisation, benign employers or the enlightened state – elements of a so-called Whig version of history whereby capitalism became progressively more humane. Rather, pointing to the struggles of organised labour, historically and contemporaneously, indicates that fluctuations in levels of workplace safety can partially, but crucially, be related to the strength of organised labour or, more accurately, the strength of balances of forces between states, capital and labour.[14]

These points being made, then, there can be no doubt that the relative decline in trade union membership and the erosion of a range of rights held by workers in the post-war liberal democracies has more recently re-shaped the balance of power between capital and labour. The internationalisation of neo-liberalism has profoundly repositioned labour markets and labour processes in favour of capital, particularly in the developing world, and this has had profound implications for the way that nation states imagine and act out their relationship to national and transnational capital. These relationships are increasingly creating and intensifying job insecurity – and all that accompanies it, including the increased likelihood of exposure to workplace risk and harm. For example, as Doogan's analysis of labour turnover in the UK in the 1990s indicates, while long-term employment actually *increased* for skilled, managerial and professional groups, job turnover increased 'absolutely and relatively in elementary occupations in lower skilled occupations and absolutely in low skilled agricultural jobs' (Doogan 2001: 432).

Such specifics are crucial in terms of the production and distribution of workplace harm. HSE research, based upon UK Labour Force Survey returns, has found significant correlations between injury rates and job tenure. Thus: the rate of injury to workers in their first six months is over double that for workers who have been with an employer for at least a year (Health and Safety Executive 2000b: 3), a correlation that holds after allowing for occupations and hours of work (Health and Safety Commission 2001: 62). Even more strikingly, higher rates of injury are associated with shorter working weeks, so that 'those working less than 16 hours per week have double the rate of injury compared with those who work 30–50 hours per week' (ibid.: 4). Thus, '[w]orkers on a low number of weekly hours have substantially higher rates of all workplace injury than those working longer hours, and the rate gets lower as the number of weekly hours increases' (ibid.: 61). Again, this holds even after allowing for other job characteristics (ibid.).

Now, there is an obvious issue here about the extent to which people learn on the job, but such facts should also raise concerns about the exacerbation of vulnerability through casualisation of work: temporary and part-time workers – workers who are likely to enjoy far fewer forms of employment protection than those in full-time, permanent employment – are, once all other factors have been controlled for, more likely to be injured at work. To these conclusions we can safely add that child workers, home-workers and those working 'illegally' are at the same time likely to face extreme risks whilst also being invisible to almost any 'official' considerations of the distribution of injury and disease. And the converse of these arguments is that the safest workplaces in the UK are those with strong trade unions and effectively functioning, union-backed safety representation. This has been confirmed in numerous research studies. What the most recent studies have found is that strong worker representation in the workplace has a markedly beneficial impact upon health and safety performance (Walters *et al.* 2005) and that trade unions are the key organisations in providing this support. Strong trade union representation reduces injury rates dramatically, by perhaps as much as 50 per cent (Reilly *et al.* 1995). Moreover, a review of the research confirmed the consistency of those findings on an international basis (James and Walters 2002). This review of evidence underlines two points. First, at the level of the workplace, the crucial risk controls are those that are exerted by well-organised workers themselves. This suggests that rather than having been made redundant in the current phase of capitalism, 'traditional' forms of labour organisation remain just as important as they have ever been. Second, it suggests that were a de-institutionalisation process to occur, it would, quite literally, have catastrophic affects. Despite the efforts of employers in particular sectors to remove negotiations around safety from the sphere of industrial relations, this strategy has only had limited success (see for example, Woolfson *et al.* 1996). Regardless of the relative decline in trade union membership, trade unions and other 'traditional' forms of worker organisation remain crucially important to the mitigation of risks in the workplace. Safety conditions remain institutionalised in collective bargaining agreements in the UK. Thus if there are clear roles for a range of campaigning groups and NGOs in seeking to intervene in safety politics,[15] these complement, rather than substitute for, 'old' forms of collectivism.

This point also helps us to explain differences in risk of death and injury in terms of sector as opposed to occupation. About half of the fatalities incurred by employees in Britain occur in two of the most casualised sectors – construction and 'agriculture, forestry and fishing' – which, along with 'recycling of scrap and waste', also have the highest injury rates to employees (Health and Safety Executive 2004b). Tables 6–9 provide some data on the distribution of various types of injury across different sectors of the economy. It is clear from these tables that the most dangerous sectors in terms of fatal and non-fatal injuries to employees are agriculture, construction and the extractive industries. In particular, the relative dangerousness of agriculture and construction is borne out strikingly in the data for fatal injuries to the self-employed, presented in Table 7. Comparing the data here with that in Table 6, we find that, on average, across all industries, being self-employed makes you twice as likely to incur a fatal injury. It should also be noted that there are no clear downward trends in any of the sectoral fatal injury rates here.

Table 6 Fatal injuries reported to all enforcing authorities by industry 1996/97–2004/05, incidence rates per 100,000: employees

	1996/97	1997/98	1998/99	1999/2000	2000/01	2001/02	2002/03	2003/04	2004/05
Agriculture, hunting, forestry and fishing	7.6	6.7	5.4	4.53	4.7	7.9	7.3	2.8	7.0
Extractive and utility supply	4.2	7.9	4.7	3.1	4.6	6.9	1.6	5.1	1.3
Manufacturing	1.3	1.3	1.6	1.0	1.2	1.3	1.2	0.8	1.3
Construction	8.2	5.7	4.4	5.5	6.5	5.3	4.9	4.3	4.8
Services	0.4	0.4	0.3	0.2	0.4	0.3	0.3	0.4	0.3
All industries	0.9	0.9	0.8	0.7	0.9	0.8	0.7	0.7	0.7

Table 7 Fatal injuries reported to all enforcing authorities by industry 1996/97–2004/05, incidence rates per 100,000: self-employed

	1996/97	1997/98	1998/99	1999/2000	2000/01	2001/02	2002/03	2003/04	2004/05
Agriculture, hunting, forestry and fishing	14.3	8.7	15.0	13.0	19	11	12.9	21.3	14.7
Extractive and utility supply	-	10.3	13.0	13.7	-	-	-	0.4	-
Manufacturing	2.3	2.7	2.2	1.2	1.8	0.4	0.5	1.2	0.4
Construction	3.0	3.1	2.8	3.2	5	2.9	2	2.5	1.7
Services	0.7	0.5	0.4	0.5	0.4	0.2	0.4	0.3	0.4
All industries	2.3	1.8	1.9	1.7	2.4	1.3	1.3	1.8	1.3

Table 8 Non-fatal major injuries reported to all enforcing authorities, incidence rates per 100,000, by industry 1996/97–2004/05: employees

	1996/97	1997/98	1998/99	1999/2000	2000/01	2001/02	2002/03	2003/04	2004/05(p)
Agriculture, hunting, forestry and fishing	256.9	222.3	205.6	224.4	213.9	238.5	262.5	235.2	228
Extractive and utility supply	315.1	282.7	246.8	244.1	267	222.9	222.7	228	232.5
Manufacturing	206.4	216.1	201.5	204.1	194.2	194.9	194.3	194.1	192.3
Construction	403.0	382.3	402.7	395.9	380.9	356.1	354.9	327.7	299.4
Services	90.8	88.4	83.7	79.3	75.3	79	80.7	94.5	93.4
All industries	127.5	127.6	121.7	116.6	110.2	110.9	111.1	120.4	117.7

p is provisional

Table 9 Non-fatal major injuries reported to all enforcing authorities, incidence rates per 100,000, by industry 1996/97–2004/05: self-employed

	1996/97	1997/98	1998/1999	1999/2000	2000/01	2001/02	2002/03	2003/04	2004/05(p)
Agriculture, hunting, forestry and fishing	40.9	32.3	36.9	41.0	34.5	54.4	43.2	42.7	41.8
Extractive and utility supply	211.7	780.8	441.5	532.7	67.2	41.4	140.7	84.8	106.4
Manufacturing	41.7	35.7	35.2	25.9	22.9	44.4	45.5	46.8	39.9
Construction	104.9	65.4	56.5	57.7	62.7	79.5	99.1	97.8	89.4
Services	13.7	4.6	5.1	5.3	4.9	8.5	9.3	12.8	13.2
All industries	38.4	23.3	20.3	19.7	19.2	27.8	32.3	33.9	32.9

P is Provisional

Table 10 Injuries to employees by occupation 2004/05(p), incidence rate per 100, 000 employees *

Occupation	Fatal	Major	Over three-day	All reported injuries
Managers and senior officials	0.3	43.0	89.2	132.5
Professional occupations	0.1	52.3	138.4	190.9
Associate professionals and technical occupations	0.3	66.9	335.3	402.5
Administrative and secretarial occupations	-	29.1	80.6	109.8
Skilled trade occupations	1.6	226.3	746.1	973.9
Personal service occupations	0.1	112.7	516.0	628.8
Sales and customer service occupations	-	83.9	316.8	400.7
Process, plant and machine operatives	3.0	408.8	1733.7	2145.4
Elementary occupations†	1.2	200.6	943.8	1145.6
All occupations	0.7	117.7	469.0	587.4

P is Provision

* Source: www.hse.gov.uk/statistics/employment/index.htm

† The elementary occupations category includes occupations non specified above in agriculture, construction, plant processing and services.

Now, in terms of relative danger across different sectors, one point needs to be made here regarding levels of 'inherent' danger, the kind of arguments that support glib assertions such as the phrase 'occupational hazard'. Whilst it is accepted that there may be greater risks to be faced on a building site than in an office, for example, the idea that certain occupations are so inherently dangerous that they necessarily produce certain levels of injury cannot be sustained. Thus within construction it is clear that there are more or less dangerous ways in which working can be organised – a crude but clear illustration of this can be witnessed in the oft-made observation that the construction work in preparation for the Sydney Olympics in 2000 produced no fatal injuries amongst construction workers, while those in Athens in 2004 produced some 40 plus (McCartney 2005). The contrast between these two contexts is generally, and plausibly, explained by, on the one hand, the strength of Australian construction unions and the legal rights they have secured in terms of workplace safety, and, on the other, the overwhelming reliance in the Athens project upon casualised, non-union, migrant labour (Australian Broadcasting Company 2004; BBC Radio 4 2004). For considerations of the organisation of danger within a different industry, the studies by both Carson (1982) and Woolfson and colleagues (1996) of work organisation in the UK offshore oil industry are instructive. Indeed, these latter studies also demonstrate that, in a sector that has drawn labour from the high unemployment areas of Scotland and Northern England, the idea that workers exchange exposure to risk for higher wages is also more myth than reality (see also Moore 1991). Thus structural vulnerability in the workplace and economic marginalisation have mutually reinforcing effects that manifest themselves in terms of exposure to workplace risks.

Gender, ethnicity and patterns of risk

While 'danger' has traditionally been associated with male occupations (Harrison 1993), and despite the fact that research has mostly ignored womens' occupational health and safety issues (Szockyi and Frank 1996: 17), trends in the data indicate that those areas in which women are over-represented, notably services, are those which exhibit both persistently high, and rising, rates of injuries and ill-health (see Craig 1981; Labour Research Department 1996). Indeed, both popular and academic understandings or considerations of the relationships between 'risk' and work views these through the prism of (declining) hazardous and male dominated occupations (mining, shipbuilding, heavier forms of manufacturing, and so on), so that much work within this area (see below) has highly, if implicit, masculinist connotations.[16] That there are both gender and ethnic both gender and ethnic dimensions to this unequal victimisation is evidenced in the work of John Wrench (Wrench 1996; Lee and Wrench 1980; Wrench and Lee 1982; see also Boris and Pruegl 1996).

An OECD report recently noted that 'Historically, work environment research has largely concerned industry work, thus the male worker' (Trades Union Congress 2002). So if there is very little social science which focuses upon occupational safety (Nichols 1997), work which does exist tends to focus predominantly upon men's work. As TGWU national organiser Diana Holland said:

> Health and safety problems are often associated with heavy industry, and jobs mainly done by men, yet accidents occur in all types of workplace and many women face hazards at work. Often it is the least obvious environments that can be most dangerous as risks go undetected and can develop into health and safety problems (Trades Union Congress 2003b).

Yet as we indicate throughout this book, occupational death and injury are routine and ubiquitous phenomena. So, if, as we noted above, 95 per cent of reported fatalities are sustained by men, and if in 2004/05 men were 2.5 times more likely to suffer a reported non-fatal injury than women (www.hse.gov.uk/statistics/tables/table11e.htm), these observations obscure a mass of relatively hidden victimisation suffered by female workers.

Thus, as Harrison has noted, if we confine ourselves to letting these overall figures 'speak for themselves', we 'miss other important issues' (1993: 258). This confirms our need to move from beyond a purely quantitative overview of occupational injuries. In the specific case of the relatively low number of injuries sustained by women, Harrison observes that this should not detract from an analysis of those industries where injuries are incurred by women since 'we would fail to appreciate the significance of injuries and disability and possible consequences for women's lives' (ibid.: 258), including particularly acute financial disadvantage (ibid.: 260).

Yet there is a prior issue here, one which reappears contemporaneously in relation to, for example, stress and musculoskeletal disorders, both of which are noted below. Each of these categories of 'ill-health' is currently a site of contestation,[17] with campaigners arguing that they are in fact matters of occupational injury/safety. This both under-scores and develops the point made in Chapter 1 that safety is not health. There is a close relationship between these different categories of detrimental effects in causation as we noted, but there is also a close relationship in their manifestations. Thus a consideration of women's safety is one way to address the problematic distinction between ill-health on the one hand, and an injury on the other – since women experience less of the latter but may experience much more of the former, even if these effects are unrecorded. For example, women constitute of 80 per cent of 'cleaners and domestics' (the *Guardian*, 28 February, 2006: 4), and are thus exposed to a whole series of chemical hazards, biological hazards and physical agents (noise, vibration) and noxious substances, yet the translation of exposures into notified cases is negligible. Indeed, the very couplet, with an implied distinction between, 'health and safety' may itself incorporate a gendered division (Harrison 1993). It has now become commonplace within trade union circles to refer to the 'gender gap' in occupational safety.

Such considerations are all the more salient given that, in Great Britain for example, women now constitute 46 per cent of the workforce (Equal Opportunities Commission 2005: 8), but many tend to work part-time – this is the case for 44 per cent of all women workers (compared with 11 per cent of men; ibid.). Further, women are three times more likely than men to be homeworkers, they tend to stay with the same employer for shorter periods, and over a third stay with an employer for less than two years. These are all characteristics of particular vulnerability. Female vulnerability at work and in the labour market is further under-scored by the fact that women remain under-paid compared to male counterparts: female full-time average hourly earnings are 19.5 per cent lower than equivalent male earnings; female part-time average hourly earnings 40 per cent lower than equivalent male full-time earnings (Trades Union Congress 2005c: 1). Moreover, vertical segregation, the under-representation of women at higher levels in occupational life, remains entrenched. According to the Government's Women and Equality Unit, women comprise 30 per cent of managers in England, while just 4 per cent of directorships are held by women; women make up 73 per cent of managers in health and social services, yet constitute over 80 per cent of the workforce; they account for just 6 per cent of managers in production.[18] Horizontal segregation also has implications for victimisation to safety crimes – if less obviously so. Women's employment remains grossly over-represented in various service occupations – 84 per cent of women employees work in the service sector (Equal Opportunities Commission 2005: 9). If women do not work in traditionally dangerous occupations, those areas of work in which they predominate are characterised by gender-specific, and less visible, risks. For us, there are clear safety correlates of this skewed distribution of poorly protected work. Moreover, women workers still experience the classic double burden, a function of their continuing to bear predominant responsibilities for child-rearing and domestic labour (Doyal with Pennell 1979: 216).

The TUC regularly surveys its safety representatives to identify what problems, caused or made worse by work, are being raised by women workers (Trades Union Congress 2002). Successive surveys have found stress to be the most frequently identified issue (ibid.). Stress affects working women more than men, and is a function of vulnerability. As the ILO has noted, 'The relation between gender, work and stress is complex and varied' (ibid.). Factors which exacerbate women's exposure to stress include low pay, repetitive, monotonous work, low autonomy and a lack of control, over-demanding jobs and greater total workload, and employers' failures to accommodate family responsibilities.

After stress, the most recent TUC survey revealed 63 per cent of women citing manual handling as the most problematic safety issue, reinforcing previous surveys 'showing that women at work are exposed to heavy lifting. Many women work in jobs where frequent manual handling is expected' (ibid.).

Third in the list are repetitive strain injuries (RSI), which are also known as work-related upper limb disorders. Fifty three per cent of the safety reps identified this category as a problem caused or made worse by work. This figure is higher than the 37 per cent who identified RSI as a problem for men and women workers in the 1998 TUC Safety Reps Survey, and reinforces the view that women are at particular risk of repetitive strain injuries: telephone operators, typists and Display Screen Equipment workers, packers, supermarket checkout operators and light assembly workers are all overwhelmingly associated with RSI. A large TUC affiliate that organises in the public sector has previously stated that

> Women are at particular risk of developing work related upper limb disorders (WRULDs) ... . Women often have to work at repetitive tasks, working long shifts or on piece rates. Women are also more likely to be put upon. Bad employers (and low wages) force women workers into not taking breaks; working longer hours in poorly designed workplaces; and not complaining. Women workers are also less likely to be able to rest once they get home: instead of recovering they often have to do the housework. (ibid.)

Women also suffer more back pain at work than men – yet they are less likely than men to take time off. A HSE poll also found women were less likely to tell their bosses they were in pain. Forty-nine per cent of female workers said they suffered from recurrent back pain, compared to 40 per cent of men. But women who suffer from back problems take an average of 10.4 days off per year, compared to their male colleagues who take 33 days sick leave per year for back pain. HSE's Elizabeth Gyngell said: 'In 2004, 205,000 women took time off work to recover from back injuries, and no doubt many more suffered in silence' (Trades Union Congress 2005d).

Musculoskeletal disorders are the most prevalent recorded form of occupational injury in Britain – over 1 million cases are reported by workers per year (www.hse.gov.uk/msd/index.htm). Incidence rates for these conditions – low back pain, joint injuries and repetitive strain injuries of various sorts – are slightly higher for women than for men (www.hse.gov.uk/statistics/causdis/musc.htm). It is of interest that these are classified as ill-health, and do not figure in injury data. Though long contested by medicine and the law, and often dismissed as the complaint of the malingerer, these injuries – to the hands, the back, the arm and the shoulder – vary from acute pain, to chronic, long-term, progressively intensifying pain, to disfigurement and crippling. A list of the popular names associated with some of these injuries says something of their association with women: 'typist's cramp, washer-woman's wrist, mother's shoulder' (Huws 1987: 50–1). If we add to that list a term that emerged in Australia in the 1970s – 'ethnic wrist' – then it is clear that these are injuries associated with poorly protected, marginalised workers.

Women also experience significant levels of inter-personal violence in the workplace. British Crime Survey estimates of violence at work – threats and assaults – indicate that men and women face similar levels of risk of being assaulted at work, but that women had a higher risk of being threatened (www.hse.gov.uk/statistics/causdis/violence.htm). And HSE's survey of self-reported working conditions reveal 'a greater proportion of females than males reported ever being physically attacked (8 per cent compared with 6 per cent) or threatened (18 per cent compared with 15 per cent) in their current job' (ibid.). While BCS data indicate that 'protective service occupations' have the greatest estimated rates of assaults and threats, these are followed by 'health and social welfare associate professionals' – which tend to be female jobs. Employers have a legal duty to risk assess to prevent violence, and this assessment must account for the specific risks faced by female workers. Compliance with and enforcement of this duty is poor (Trades Union Congress 2002).

Given the peculiar nature of women's vulnerability to certain kinds of injuries, the legal requirements for employers to assess risks should often include a gendered element. However, recent survey evidence indicates the failure of employers to meet their legal duties in this respect. Thus, of employers' responses to problems raised by women workers, this evidence concludes that: 26 per cent of employers always take the problems seriously; 56 per cent of employers sometimes take them seriously; 12 per cent of employers rarely take them seriously; and 3 per cent of employers never take them seriously (ibid.). Thus, only in one out of four cases does management always take the problems raised by women workers seriously. The particular health and safety concerns of working women need to be effectively addressed and taken seriously on all occasions by all employers. Twenty seven per cent of risk assessments have not addressed women's concerns, and 41 per cent have only partly addressed them. So in less than three out of ten cases are employers fully addressing the risks faced by women workers despite the legal requirement to do so (ibid.). This is as likely to be borne of an implicit assumption of 'the worker' as being male rather than out of any conscious discrimination; but whatever the reason, the effects are the same. Thus, for example where women and men are performing the same tasks, women could be placed at greater risk of injury than men because the job equipment and personal safety equipment are designed to fit the average man (Smith and Mustard 2004). In its respect, one could conclude that the relative invisibility of women's safety issues is exacerbated by employers' safety compliance efforts, in turn reproducing women's invisible victimisation.

The association between vulnerability and victimisation is further highlighted in recent HSE-sponsored research examining the experiences of different ethnic groups of work-related stress (Smith *et al.* 2005). Nationally, HSE estimates that about half a million people in the UK experience work-related stress at a level they believe is making them ill, with a total of 12.8 million working days lost to stress, depression and anxiety in 2003/4. Stress, of course, has multiple causes, but is generally experienced more greatly by those who are least able to exert any control or influence over workplace demands placed upon them. It is, then, a classic disease of workplace marginalisation and vulnerability. Smith *et al.* found a 'significant correlation' between work stress and ethnicity. In summary, they concluded that '[r]acial discrimination, particularly in combination with gender and ethnicity, was identified as having a strong influence on work stress . . . certain work characteristics were also associated with work stress: higher effort reward imbalance, greater job demand, and lower control over work were all associated with work stress' (ibid.: viii). Put simply, despite popular myths about stress being associated with high flying corporate executives, if you are a minority-ethnic female, low down the occupational hierarchy with little or no control over your work, and both over-worked and under-paid, then you are likely to be subject to this form of occupational disease. Understanding exposure to and victimisation from risk, and differential responses to and public knowledges of this victimisation, thus requires an understanding of various forms of class–race–gender articulations. Such an understanding requires deconstructing traditional representations of workplace risk and hazards.

Globally, a particularly heavy toll of death and injury occurs in developing countries where large numbers of workers are concentrated in primary and extraction activities such as agriculture, logging, fishing and mining (Takala 2002: 2–3). This unequal international distribution of risk is further exacerbated by the struggle to secure safer and healthier workplaces in developed countries having as one effect the relocation of risk and hazard to developing economies – creating a key advantage for Transnational Companies in their ability to 'export' hazardous work from more to less regulated contexts economies (Castleman 1979; Ives 1985). The classic and most infamous manifestation of this was the Bhopal 'disaster'. While the gas leak was the result of a combination of production and cost-cutting decisions (Pearce and Tombs 1998), the very fact that MIC was on-site at all in Bhopal was the consequence of successful resistance by French chemical workers to a prior Union Carbide intention to store and use MIC at its plant in Beziers, France (ibid.).

The long and still largely unsuccessful struggle of the victims of the Bhopal gas leak to gain compensation for human and environmental suffering also indicates a further structural inequality associated with global risk distribution: for the economic consequences of death, injury and disease associated with working are also differentially distributed. Thus, for example, occupational health and safety compensation schemes differ enormously – while workers in Nordic countries enjoy nearly universal coverage, 'only 10 per cent or less of the workforce in many developing countries is likely to benefit from any sort of coverage. Even in many developed countries, coverage against occupational injury and illness may extend to only half the workforce' (Takala, cited in Demaret and Khalef 2004).

Finally, to talk of deaths and injuries is not merely to refer to *physical* harms. Occupational injuries – however 'minor' – have widespread, if largely unrecognised, financial, psychological, as well as social effects. However, it is the financial costs of injuries and ill health which have been the focus of recent attention by official organisations and interest groups. Thus the HSE and governments have, for over a decade now, sought (erroneously – see Cutler and James 1996) to argue the 'business case' for improved health and safety (Health and Safety Executive 1993; Davies and Teasdale 1994; Health and Safety Commission/Department of Environment, Transport and the Regions 2000), seeing the costs to companies of injuring and causing illness as a lever to raise standards of compliance! In this context, it has been estimated that the costs of injury and ill health is £18 billion a year (see, for example, Health and Safety Commission/Department of Environment, Transport and the Regions 2000, *passim*). Yet one of the contradictions within such an argument is that employers *do not* actually meet the costs of workplace injuries and illness: most of the £18 billion cost of workplace injury and illness is paid for by the government and the victims. Indeed, notwithstanding the contemporary prevalence of claims about a 'compensation culture', currently fewer than one in ten workers entitled to compensation for a work-related injury or disease actually receive this (Pickvance 2005). Even the HSE itself estimates that employers, who cause the health and safety risks, pay between £3.3 billion and £6.5 billion (Trades Union Congress 2003c). In other words, costs associated with injuries and ill health represent a socialisation of the costs of private production, in effect a massive redistribution of wealth from the poor to the rich. That is, through supporting the cost of industrial injury benefit, health and other social services, paying higher insurance premiums, paying higher prices for goods and services so that employers can recoup the costs of downtime, retraining, the replacement of plants and so on, private industry is subsidised on a massive scale by employees, taxpayers and the general public. This in turn has knock-on effects on governments' ability to use revenue from general taxation for more socially productive purposes.

Such bald statistics of deaths or losses to corporate profits or GDP mask a wide array of searing, but less quantifiable, social and psychological harms. Families and communities are subjected to trauma in the event of death and injury. Children lose mothers and fathers, spouses lose partners, sports teams lose coaches and players, church and social clubs lose their members, and workers lose their colleagues – to the extent that such psychological and social costs are immeasurable. Moreover, such losses and harms have effects across generations, so that, for example, children who experience poverty following the death of the main wage earner are themselves more likely to grow up in conditions of relative insecurity, a condition then more likely to be experienced by their own offspring.[19] It remains to be noted that the psychological trauma heaped upon the bereaved is often magnified greatly by the consistent inability of the state, through the criminal justice system, to provide 'answers' as to why someone who leaves for work either does not return, or returns in a considerably less fit condition – despite central elements to law's own stated 'promise' being, first, to provide mechanisms for bringing the perpetrators of illegal harms to account, and, second, indeed increasingly, to provide redress for victims of such harms (Tombs 2004).

Conclusion

The data that has been reviewed in this chapter indicate that no matter how incomplete the officially recorded deaths and injuries caused by work, the toll is undoubtedly huge. The reconstruction of UK fatality data presented here produces a minimal estimate of 1,700 people killed annually. However, this exercise still tells us relatively little about how victimisation is distributed. In a global or even European context, as we have pointed out, the UK may not represent an exemplar case, but it is by no means the most dangerous country in which to work. The annual death toll of people caused by work globally rivals the death toll caused by wars and dwarfs the death toll caused by acts of terrorism, no matter how broadly 'terrorism' is defined (International Labour Organisation 2005).

To the inadequacies of the headline figures on deaths and injuries, we can add major inadequacies in our knowledge of how deaths and injuries at work are distributed. Yet our review of the limited conclusions that we can draw from existing data and the few empirical studies that exist on the distribution of victimisation provides a powerful illustration of the extent to which position in the workplace, socio-economic group, gender and ethnicity structure the risk of being injured or killed. Moreover, when it comes to absorbing the social and economic costs of deaths and injuries at work, it is the most vulnerable that bear the greatest burden. None of this is revealed in any of the officially published data sets; precious little is revealed in government-funded research into the subject.

More meaningfully, this discussion has urged us to explore how exposure to risks through work produces a highly concrete manifestation of what has been called the structuring of vulnerability (Nichols 1986) – it is one of the means by which the poorest, most disadvantaged people in our society are further, systematically discriminated against in terms of the quality, longevity, even preservation of their lives. At the same time, as is often the case with structured inequalities, these exposures remain relatively invisible, further exacerbating social, economic and political marginalisation of those who face them.

Notes

1 Further details can be found at International Labour Organisation (2005) and at www.corporateaccountability.org/internationsl/deaths/tables/summary/main.htm

2 Source: www.europa.eu.int/rapid/pressReleasesAction.do?reference=STAT/04/55&format=HTML&aged=0&language=EN&guiLanguage=en. See Eurostat, Population and Social Conditions, 'Work and health in the EU – A statistical portrait'. Eurostat is the Statistical Office of the European Communities.

3 HSE note that examples of major injuries include 'fractures (except to fingers, thumbs or toes), amputations, dislocations (of shoulder, hip, knee, spine) and other injuries leading to resuscitation or 24-hour admittance to hospital' (source www.hse.gov.uk/statistics/sources.htm).

4 Moreover, even this recognised level of under-reporting still in principle excludes three potentially significant – if somewhat overlapping – areas of occupational injuries, namely those incurred by workers in the illegal economy, as well as by home- and child-workers (see, for example, O'Donnell and White 1998, 1999).

5 In the supplementing of officially-recorded data with self-report data via the Labour Force Survey, there is some similarity with the Home Office's development of the British Crime Survey as a self-report, victimisation survey aimed to overcome some of the problems associated with recorded crime statistics and the so-called hidden figure of crime; it is clear, however, that the two exercises are very differently resourced, and attract very different levels of political commitment. It should also be noted that in 2005, the HSE conducted its first Workplace Health and Safety Survey, completed by 996 employer respondents, aiming to 'record health and safety conditions across British workplaces'. One aim of this is to supplement recorded injury data. A worker survey is planned for 2006 (www.hse.gov.uk/statistics/ pdf/whassnotes.pdf).

6 This calculation is made upon the basis of a combination of reported injuries under RIDDOR and self-reported injury through the Labour Force Survey.

7 Though not apparent from the data in the table, over 95 per cent of fatal injuries to workers in any one year are overwhelmingly sustained by men (Health and Safety Commission 2001: 56–7).

8 As is just as likely to be the case with the one hundred plus annual fatalities to members of the public that are not suicides on railways.

9 Moreover, self-employment has grown dramatically in recent years, rising from just over 2 million workers in 1974 to a current total of around 3.8 million (www.hse.gov.uk/statistics/empolyment/index.htm).

10 The numbers of fatalities relating to the supply and use of flammable gas are hardly irrelevant in the context of the overall total of fatalities. In the five reporting years between 2000 and 2005, 33, 26, 25, 18 and 23 such deaths respectively were reported to HSE (www.hse.gov.uk/statistics/tables/tablegs1.htm).

11 Also excluded are injuries reportable under separate merchant shipping, civil aviation and air navigation legislation; injuries to members of the armed forces; and fatal injuries to the self-employed arising out of accidents at premises which the injured person either owns or occupies (source: www.hse.gov.uk/statistics/sources.htm).

12 For a more detailed treatment of the reconstruction of fatality data, see Tombs 1999a.
13 Thanks to Gary Slapper for first bringing these points to our attention.
14 We shall return to these arguments regarding what Snider (1991) has called the dynamics of regulatory reform in Chapter 7.
15 For example, the Simon Jones memorial campaign (www.simonjones.org.uk) or the Centre for Corporate Accountability (www.corporateaccountability.org).
16 We acknowledge both that there are exceptions to this criticism, and that the criticism applies to our own work.
17 For example, the first campaigning aim of the UK National Stress Network, part of the Hazards Campaign, is for 'the recognition of stress-related illness as an industrial injury' (UK National Work Stress Network 2006: 3).
18 Source: www.womenandequalityunit.gov.uk/women–work/women_top.htm
19 Thanks to Paddy Hillyard for first pointing out to us this particular dimension of the harms associated with occupational injury and ill health.

Chapter 3

Obscuring safety crimes

Introduction

In previous chapters – through a brief consideration of a series of case studies within the rubric of safety crimes, and then through presenting a range of data regarding the scale and distribution of occupational deaths and injuries – we have indicated that in objective, if sheer physical, terms, deaths and injuries at work represent a significant social problem. Of course, as we have emphasised, this does not necessarily mean that this social problem is recognised popularly nor politically: that is, such injuries and deaths may remain relatively invisible, or if visible, be perceived as neither a significant social nor indeed a crime problem; and each of these is in fact the case. The anomaly apparent here – a significant toll of death and injury which seems to barely register as a social or crime problem – needs further consideration. In short, how is it that safety crimes are obscured from view?

The very fact that these deaths and injuries tend *not* to be counted as crimes within official data is one in a series of intimately linked social processes through which their existence and nature is obscured and distorted. Law and legal processes play a key role in rendering such events as separate from real crime, and we consider these in subsequent chapters. Here, however, we note some of the general processes related to their more general invisibility from crime, law and order discourses, before focusing in particular upon a key element of this invisibility, namely the ways in which safety crimes are represented both as accidents, and then as phenomena that imply various degrees of culpability on the part of victims.

The invisibility of safety crimes

At the broadest level, safety crimes remain socially, politically, and academically invisible in ways which mirror the invisibility of corporate crimes in general. Thus there is an array of social processes that contribute to removing such offences from dominant definitions of 'crime, law and order' (Slapper and Tombs 1999).

Here, what has been increasingly recognised as 'the politics of law and order' is clearly important (Hale 2004; Downes and Morgan 2002; Brake and Hale 1992). Generally, as law and order has become a politicised area, then law and order discourses have drawn various events and processes *onto* the terrain of crime, expanding the reach of the criminal justice system. But this 'net-widening' (Cohen 1985) has not extended to safety crimes. Here, the effect has been quite different: in simply not considering occupational deaths and injuries within crime, law and order discourses, the difference between the former and real crime has been further cemented. More generally, beyond this rhetorical level, there are a whole series of ways in which political processes exclude safety crimes from considerations of crime, law and order.

Firstly, it is clearly Governments that formally define crime, that pass or block legislation, that fund enforcement bodies, that send formal and informal signals to the ways in which laws are to be enforced, and so on. These considerations, to the extent that they deal with the passage and implementation of the law, will be dealt with in a later chapter, where we consider the myriad ways in which law *separates out* safety offences from 'real' crimes (Chapter 5). But here we might also add some further – perhaps disparate – observations which point to the de-criminalisation of safety offences through political processes. Institutionally, the Government department that deals with crime is the Home Office; the office of state that covers safety regulation is (currently) the Department of Work and Pensions – not exactly synonymous with crime-fighting. Running the Home Office – that is, being Home Secretary – is now one of the highest profile Government offices, ranking close to Chancellor and Foreign Secretary in seniority. Responsibility for health and safety, by contrast, is located at junior minister level, within the DWP. A further contrast is found in the level of political stability afforded to these different areas. From the election of the first Blair Government in 1997, to 2007, three Labour Governments have seen just four ministers running the Home Office (Straw, Blunkett, Clarke and Reid)[1]; in 2007, Lord McKenzie became the seventh minister responsible for safety and health at work in seven years, and the ninth since Labour came to power in 1997. Long before Lord McKenzie's appointment, unions, safety campaigners, RoSPA and even the Institute of Directors bemoaned the lack of ministerial continuity, prompting the hardly radical Institution of Occupational Safety and Health to note how this game of 'pass the safety parcel' . . . 'begs the question of how seriously the government takes safety, when it still hasn't delivered on its manifesto commitments' (Jolliffe 2004: 2).[2]

Of course, a key product of the lack of political priority attached to safety offending is the very poverty and paucity of data in this area which was the focus of the previous chapter. If, as Maguire notes, 'a salient feature of almost all modern forms of discourse about crime is the emphasis placed upon terms associated with its quantification and measurement' (Maguire 1994: 236), then it is hardly surprising that neither safety crime in particular nor corporate crimes in general 'feature in . . . debates about the "crime problem"' (Nelken 1994: 355; Green 1990: 27).

This separation of safety crimes from the proper sphere of crime, law and order politics is reflected in, and in turn reinforced by, the sheer torrent of criminal justice legislation aimed at the usual 'suspects'. Thus it has been estimated that the period of Labour Governments since 1997 has seen over 50 pieces of crime and immigration legislation pass through Parliament (Hillyard 2006), while it has been generally claimed that over 1,000 new criminal offences had been created up to May 2005.[3] None of these of course has impinged upon safety crimes – with the one major piece of legislation in this area, new legislation to make it easier to prosecute companies for corporate manslaughter, taking ten years from the point at which it first appeared in Labour's manifesto to the point at which it appeared in Parliament for debate (see also Chapter 6).

If we leave the sphere of formal politics, it is also clear that in any case long-standing constructions of crime and the criminal create obstacles to businesses and business people being viewed in similar ways to conventional offenders: thus there is also a significant lack of fit between offenders in the sphere of safety crimes, and those in the context of conventional crimes. In terms of business organisations, corporations may have been granted the status of legal persons, but they do not equate with popular images of real persons, let alone 'criminal' ones; similarly, directors of and managers within them simply do not fit our popular images of offenders (see Lacey 1995: 21). Individual business offenders simply do not 'look the criminal type'. Although any first year student of criminology can articulate a series of critiques of Lombrosian categories, the idea that criminals look a certain way – or at least do *not* look certain ways – is, in our view, enormously powerful. When the photographs of six men (four from the former line operator Railtrack and two from the construction firm Balfour Beatty) appeared on the front of national and regional newspapers on the day that their trial for manslaughter charges following the Hatfield train crash commenced (Tuesday February 1, 2005), it was immediately apparent that *they simply did not look like killers.*

Moreover, if it has *historically* jarred to think of businesses and business people as offenders, this has become even more the case under the emergence to dominance of neo-liberalism, where the legitimacy afforded to businesses and business people has been significantly augmented (Snider 2000; Tombs 2001). Indeed, more recently, far from being potential criminals, business people are those whom we are supposed increasingly to aspire to be, role models of synoptic societies, and even – as in the case of some like Branson, Sugar and Trump – celebrities. Corporations and those who work within them, then, are viewed quite differently to the objects of 'traditional' crime concerns; 'conventional criminals' tend to be represented as a burden upon society in a way that corporations will not be. Further, where business organisations engage in criminal activity, this tends to be cast as involving technical infringements of law, rather than real crimes. Finally, and crucially, as we have argued elsewhere with respect to corporate crimes in general, offences are represented as an aberration from their routine, legitimate activities – a point we shall take up below with respect to the power of the language of accidents.

These assumptions, and the general contrast with 'real' crime and 'real' criminals, are reflected in and reinforced by the media. Whether we survey fictional or documentary-style treatments of crime on TV, or newspaper and other print media coverage of the issues, we find that while there may be some attention to safety crime – specialist documentary programmes, for example – representations of crime converge to produce 'blanket' conceptualisations regarding 'law-and-order' that reinforce dominant stereotypes of crime and the criminal (Chibnall 1977). Thus, as with corporate crimes in general, where safety crime is covered, its presence is vastly outweighed by treatments of conventional crime, it is treated in lower profile outlets or formats, and is often represented in the rather sanitising language of scandals, disasters, abuses and accidents rather than as criminal activity (Tombs and Whyte 2001).

The point here is not, of course, to claim that the media creates our images, ways of seeing and understanding: the situation is much more complex than that. But one of the reasons why the media is particularly powerful in the context of corporate crimes in general and safety crimes in particular, is because these already seem for many of us to be far removed from our immediate experience. And this is partly an effect of the relationship between the offender and the victim in many corporate offences. In most forms of traditional crime there is, or must at some point be, a degree of proximity between offender and victim. By contrast, with safety crimes, there are frequently enormous distances between offender and victim, in terms of both space and time – many of the operating conditions on Piper Alpha or at Bhopal or the ship on which Simon Jones was working were determined many miles away in the United States or Holland, by 'faceless' managements in Union Carbide or Occidental Petroleum or Euromin Limited., who had no contact, *let alone any relationship*, with the workers or passengers who were to die as a result of decisions taken – or not taken – some time in the past (see Chapter 1).

Indeed, many victims of both corporate and safety crimes may never be aware of their status as victims (Croall 2001; Meier and Short 1995). A response to sustaining an injury at work may be to chide ourselves to be more careful in the future – we are at least as likely to blame ourselves (below) rather than think that we may be the victim of an offence. And even if people are, or become, aware of their status as victims, acting upon this awareness is often extremely difficult. While it is an issue of health rather than of safety, if we return to asbestos-caused diseases, this much is clear. Thus, the asbestos-industry tried to suppress information about ill-health effects for over 100 years by co-opting the medical community, attacking critical science and funding industry-friendly research, by forming apparently 'independent' lobbying groups, and through quite simple and routine cover-ups, manipulation of data, and lying to workers and regulators; and then Governments failed to support victims in their efforts to secure compensation.

Finally, even if aware of their victimisation, it may be perfectly rational for injured workers or their bereaved families *not* to act on this – for example, to attempt to pursue the issue of any criminal offence having been committed. For one thing, HSE and its inspectorates, while formally charged with responding to complaints, are notoriously reluctant to be seen to take the side of workers since they have been adamant since their formation that they do not become involved in 'industrial relations issues' – and, as we shall see, operate on the basis of there being a natural 'identity of interests' between employers and employees on the issue of safety at work. Thus the chances of 'success' of working through formal channels are slim indeed. And against this need to be balanced the reactions of employers to those raising safety complaints – which can range from intimidation through to barring an employee from a whole industry (see Tombs and Whyte 1998). Finally, the widespread focus upon behaviour-based safety schemes (see below) and their links to 'no-accident' bonus schemes within workplaces and companies acts as another powerful disincentive against acting upon one's recognition of possible victimisation.

Now, none of the various mechanisms whereby safety crimes are rendered relatively invisible are particularly remarkable in isolation. What is crucial, however, *is their mutually reinforcing nature* – that is, they all work in the same direction and to the same effect, removing these crimes from 'crime, law and order' agendas. Moreover, they operate alongside a powerful set of discourses which create injuries within the context of the workplace as something always/already distinct from crimes – that is, as *accidents*.

Occupational *accidents*

It remains common, indeed ubiquitous, to refer to incidents that lead to death or injury at work as 'accidents'. This term is a neutral, 'anaesthetising' one; by contrast, alternative terms, such as industrial killing, wounding, or violence, would carry with them quite different connotations and logically be related to quite different legal, political and social responses. The language of accidents infects almost all statements and documents on the issue of workplace deaths and injuries, on the part of Governments, regulators, NGOs, academics and even trade unions and campaigning groups. Thus, for example, in his Foreword to the Government's flagship strategy document, *Revitalising Health and Safety*, Deputy Prime Minister Prescott refers to the 'tragic rail accident' at Ladbroke Grove – an incident still subject to a criminal legal investigation at that time[4] (Health and Safety Commission/DETR 2000: Foreword, and *passim*). A major HSE report on 30 years of health and safety legislation and enforcement presented a self-congratulatory table headed 'The Decline in Fatal Accidents', despite the fact that HSE does not record *accidents*, only fatal injuries (Health and Safety Executive 2004c: 5) – a nomenclature repeated by academics when they speak of this area, even when they claim awareness of some of its problematic implications (see Hutter 2001: 50, *passim*). As we noted at the outset to Chapter 2, the ILO has recorded how the number of 'fatal occupational accidents' is increasing internationally (International Labour Organisation 2005: 1).

Now, these observations are not raised in any point-scoring fashion. The point is that all of those who comment upon safety incidents in the workplace tend to fall easily at times into using the language of accidents – including ourselves, we are sure – and this indicates its ubiquitousness, its pervasiveness, its all-too-easy commonsensical appeal. The language of accidents is both socially available and socially prioritised. And it is for these reasons, and as a preface to our discussion here, we also need to make clear that this language is not generally used consciously. At the same time, though, we must recognise that at times use of this language *is* the result of conscious corporate manipulation, both generally and in relation to specific events (Wells 1993: 40). As Wells notes, and as is particularly of relevance given the nature of victimisation in the case of many forms of safety crimes, the further removed the harms they cause are from the public experience, then the easier is such manipulation (ibid.: 40).

But of more interest to us are the ways in which the use of particular forms of language need not be understood simply, or perhaps even largely, as the result of conscious manipulation, since 'The social construction of behaviour and events results from a complex interaction between a number of factors, including cultural predispositions, media representations, and legal rules, decisions, and pronouncements' (Wells 2001: 11). That is, we need to know something of how this language is sustained and, further, what its power and effects are.

One observation that needs to be made is that the applicability of the term accident can be sustained partly due to the low level of investigation of industrial injuries, near-misses, and so on, as we shall see in Chapter 4. This systematic lack of analysis of the causes of workplace incidents is one of the key mechanisms by which the ideology of the accident can persist, both for the very lack of scrutiny of actual causes and also because the fact of an investigation immediately raises a particular incident as unusual, and thus anomalous, or, in Mathiesen's (2004: 37–48) terms, 'isolated'. In mutually reinforcing fashion, of course, the lack of investigation itself is partly based upon the idea of the 'accident': in their study on the police investigation of safety crimes, Alvesalo and Jauhiainen (2006) found how incidents were often described and defined as accidents, with the contributory negligence of the victim emphasised; such constructions influenced police work, and were used as reasons *not* to proceed with a crime investigation.

In our view, then, the predominance of the language of accidents (and associated terms like tragedy) – along with derivatives reserved for multiple fatality incidents such as disaster – to refer to workplace deaths and injuries carries with it several, reinforcing, sets of implications, each of which even in isolation renders it less rather than more likely that the events and processes thus obscured can be viewed as crimes:

First, the term 'accident' carries with it implications regarding intentionality, or the lack of it, which are crucial in the context of safety and health crimes. Thus Goldman notes that what she calls 'accidentality terms',

> provide us with an account of the mental element – intention, will, desire, deliberation, purpose, etc. – in some event. When describing some incident or process as an 'accident', or having 'occurred/been done accidentally' rather than 'deliberately' or 'intentionally', we conflate information not just about causation, and perhaps (if pertinent) degrees of culpability and fault to be imputed, but also about the element of consciousness that intruded into the event (Goldman 1994: 51–2).

As we shall demonstrate in Chapters 5 and 6, the concept of intent enjoys significant legal status – and, relatedly, for some it is a key distinguishing criterion between conventional ('real') crime and corporate crime (merely technical offences). Below, in this chapter, we shall also set out how it is in fact problematic when set against common-sense standards of moral seriousness. For now, it is enough to say that, given that intent is deemed intrinsic to crime, or at least to serious crime, then to ascribe to deaths and injuries a status implying a lack of intent, that is, 'accidental', is immediately to obscure an inter-related set of issues around seriousness, criminality, fault, causation, and so on.

Second, the language of 'accidents' is one which focuses upon specific events, abstracting them from a more comprehensible context (Scheppele 1991). It evokes discrete, isolated and random events, processes, systems and carries with it connotations of the unforeseeable, unknowable and unpreventable, despite the fact that any examination of a range of incidents reveals common, systematic, foreseeable and eminently preventable causes and consistent locations of responsibility. Such language is one way of performing the techniques of isolation referred to above, following Mathiesen, whilst it also already helps to pre-empt further, and certainly critical, scrutiny – or, scrutiny of certain kinds. Thus 'If accidents are discrete, unforeseen and chance events . . . then it becomes difficult to see them as social events, as socially contexted' (Harrison 1993: 255).

In these uses of apparently innocent, or neutral, linguistic terms which in fact carry enormous social and political meaning, there are striking parallels here with Mathiesen's recent, and brilliant, analysis of political repression in advanced capitalist economies. Whilst accepting that in such efforts physical coercion still has key roles to play, Mathiesen emphasises 'silent silencing'. Silencing is the creation of acquiescence to dominant ways of understanding the world:

> Silence in this sense is a continuum, from silence despite disagreement (grudgingly you go along) to silence as an accepting attitude (you accept the standpoint, not even noticing that silencing has taken place, or at least not taking the fact of silencing seriously). (Mathiesen 2004: 9)

The silent element of the silencing process refers to the creation of acquiescence 'through a process which is quiet rather than noisy, hidden rather than seen, non-physical rather than physical' (ibid.).

And these are precisely the ways in which the processes pointed to above operate. Indeed, a key illustration of silent silencing for Mathiesen is afforded by the capsizing of the Alexander Kielland, an oil installation in the Norwegian North Sea, in 1980, which resulted in 123 deaths. For Mathiesen, the scale of this event was one which posed fundamental questions regarding the extraction of oil from the North Sea, that is, as an event it became a challenge to authority and to a key economic activity. To some extent, then, the incident is similar to the fire and explosion on Piper Alpha in July 1988 (a threat to oil production and thus to the British economy), the capsizing of the *Herald of Free Enterprise* (a threat to passenger ferry services within an increasingly competitive market, and to a large conglomerate, P&O), the gas leak at Bhopal (which put into focus the activities of chemicals companies across the developed and developing worlds), or the series of train crashes that beset the UK in the 1980s and 1990s (which, by the late 1990s, finally threw the network into chaos, this leading eventually to the demise of Railtrack). In the case of the Alexander Kielland, such was the threat posed, claims Mathiesen, that a concerted response was required:

> because people's perception of totality or context is dangerous to an activity like oil extraction it becomes important for the representatives of the activity to pulverise the relationships which people begin to see (ibid.: 37).

Silencing through pulverisation – which becomes 'more important the more extensive and sensational the event is' (ibid.) – requires a series of techniques to be deployed, and for us these are useful when thinking about the ways in which power deals with the mundane, routine industrial deaths and injuries and the acute events that erupt into public consciousness from time to time.

Most crucially, the event in question needs to be isolated. Pulverisation through isolation encompasses a series of techniques in the form of messages through mass media and political discourse. Of particular interest in the context of this discussion is that for Mathiesen this firstly entails the event being 'individualised', that is, the event is 'made into something unique, something incomparable, and something quite special, individual and atypical', far too exceptional for any generalised lessons to be drawn or arguments made (ibid.: 38). The event may also be 'normalised'. Although this may seem contrary to individualisation, it is perfectly comprehensible as a complementary technique: for normalisation can only proceed effectively if the event is pulled out of a regular, familiar context and placed into another; and within this other context – the high risk industry or the inherently dangerous activity or, we would add here, the at–risk worker (below) – then this apparently unique event becomes entirely usual, relatively typical and ordinary. The event may also be 'split up', by which it is meant that the event is reduced to a series of loosely connected, relatively minor details of an incident – often technical in nature – which are thereby torn from their systemic context and their systematic relationship with this context; thus post-Piper discussions could come to revolve endlessly around by the availability of Temporary Safe Refuges, post-Zeebrugge by technical questions of ro-ro ferry design, or train safety by the cost and merits of Automatic Train Protection systems.

Now these arguments are important in themselves. But what is also crucial about them, for us, is how Mathiesen's claims regarding isolation are crystallised through the language which sits at the heart of discussion – official, political, academic or popular – of occupational safety, namely the language of accidents. And it is to a further, more specific consideration of the nature and power of such language that we now turn.

Victim blaming and the accident-prone worker

If the language of accidents carries with it a whole series of connotations, there is one further set of consequences to be discussed here, in some detail, which revolve around the issue of victim blaming. The basis of victim-blaming discourses is to be found in the idea of 'accident-proneness' which, in the context of occupational safety, has a very specific history. Sass and Crook (1981) note that the term 'accident prone' was first coined in 1926,[5] when studies conducted in the UK concluded that 'varying individual susceptibility to "accidents" is an extremely important factor in determining the [accident frequency] distribution' (ibid.: 183). Latterly, Grayson and Goddard set out its meaning:

> Accident proneness is generally used to mean that accidents are caused by a few individuals who tend to have a large number of accidents – and that they have repeated accidents *no matter what precautions the employer takes* (Grayson and Goddard 1976: 17, emphasis added).

According to these definitions, then, accident proneness is a 'quality' of individuals, something associated with personality types, personal characteristics and/or lifestyles, and a 'quality' that they bring with them into the workplace. And it is with the emergence of this notion of accident proneness that we can perceive some kind of foundation or basis for the prevalence of the plethora of arguments (and tactics) that can be subsumed under the practice of worker/victim blaming. We also see the basis for a whole series of professional/'expert' interventions into the workplace which address not dangerous work organisation or plant, but which attempt to modify the behaviour of workers. In this, there is perhaps an analogy to be drawn with the 'birth' of criminology, as documented by Foucault, this taking on a particular form on the basis of a specific definition of the problem to be treated:

> [An individual] steals because he is poor, certainly, but we all know that all the poor don't steal. So, for that individual to steal, then there must be something wrong with him. This something is his character, his psyche, his education, his unconscious, his desires. And with that the delinquent is handed over to a penal technology, that of the prison, and a medical technology, if not of the asylum, at any rate that of specialised supervision (Foucault 1977 : 12).

Thus, as not all workers are involved in accidents, despite working in the same environment as colleagues, then some of them are defined as accident-prone (or a variation upon this – lazy, stupid, careless, and so on) just as the poor who steal are defined as 'criminal' or 'criminogenic', since not all the poor steal. The fault is not in the system which maintains the poor in poverty, just as it is not in the working environment which presents hazards to a workforce.

Fifty years after the 'discovery' of accident-proneness in studies by Greenwood and others, the Robens Committee – whose report largely determined the nature and substance of the 1974 Health and Safety at Work Act – itself explicitly recommended that government address this problem in its approach to health and safety regulation. According to the particular version of this concept contained in the Committee's Report, the single most important reason for accidents at work was worker apathy (Robens 1972: 1). Asserting that 'Safety is mainly a matter of the day-to-day attitudes and reactions of the individual', it concluded that accident prevention should focus upon the fostering of safety awareness (ibid.: 1-2). Thus the Robens Committee pinpointed those workers involved in, or close to the scene of, industrial accidents, rather than unhealthy and unsafe working conditions, as the locus of potential improvements in industrial health and safety.

There is no intention here to enter into a critical commentary upon the original studies conducted by Greenwood and others on the distribution of accidents among a particular cohort of workers – this has been done more than adequately elsewhere (Sass and Crook 1981). What is important for our purposes is to note the shift in the use of, or reference to, accident-proneness. This 'concept' is no longer used in any specific sense, as originally developed – that is, referring to workers who have been 'shown' statistically to be more likely than others to have accidents. Rather, it has come to be used in a much more generalised, vague but highly functional fashion – workers as a whole are viewed as a group of people predisposed towards accidents, through incompetence, carelessness, apathy, recklessness, and so on. Accidents are linked with workers. Those on the scene of accidents are held, almost automatically, as responsible. In a generalised sense, then, 'workers' become the problem at the heart of occupational safety – and this in turn is the basis for worker, and often victim, blaming in discourse and practice around occupational safety (and, moreover, this set of tactics is one amongst a variety of strategies by which the victims of social problems in general are held responsible for their plight – see Ryan 1971; Goodey 2005: 96–9).

Where a worker is blamed (either explicitly or implicitly) in practice, this tactic will at times involve resort to one or several characteristics – that they are incompetent, lazy, apathetic, inattentive, careless, reckless, and so on. Crucially, although the notion of risk is directly invoked, it is seen not as something that exists apart from, and thereby imposing itself upon, the worker; rather, risk is viewed as something which originates in / is brought on by the behaviour of workers. Thus the most obvious consequence of accident-proneness and worker/victim blaming is the diversion of attention from the fact that risks are built into the very system of capitalist production, and subsequent lack of emphasis upon the nature of the workplace, the processes involved therein, and their particular rationales. If the focus on the aftermath of an incident were not to be upon workers but shifted to employers – to their employment practices (levels of training, or pressures in terms of the speed or intensity of work, for example), or to the design of the workplace and/or the job – such attention may logically lead to either cost implications, to redress the lack of investment in plant or people, or have legal ramifications, or both. As Robertson notes, 'The use of the word "accident" to describe when a worker is injured or killed hides the fact that often this is a result of a criminal act by the employer . . . . Words are being used to change the way we deal with health and safety' (2004: 6).

This strategy has a long history. Doran has argued that the very concept of an industrial accident was 'reorganised' by official discourse through the early Factory Acts, so that these events labelled accidents became separated from workers' own experiences of the broader problem of their health (understood within the organisation of production), and became subject to routine interpretation (observation, recoding, codifying) by factory inspectors (Doran 1996; see also Watterson 1991). This reorganisation itself created the possibility for the emergence of the statistically recorded unavoidable accident, subsequently situated within discourses of risk and insurance rather than negligence, culpability or criminality (Doran 1996; Defert 1991; Ewald 1991).

More generally in the context of occupational safety the idea of accident-proneness has been deployed quite explicitly also to invoke events or phenomena in which victims are implicated, via their carelessness, apathy, or lifestyles (alcohol or drug abuse, poor sleeping habits, and so on; Tombs 1991). That is, victims become complicit in their victimisation –

so that they are either to be blamed or at least not to be viewed as innocent.[6] Thus Robertson reports how, in 2000, the *British Medical Journal* reported that it would no longer use the word 'accident' in its pages, since this was 'inappropriate': 'it argued that "most injuries and their precipitating events are predictable and preventable". It added that by using the word "accident" we were encouraging victim-blaming and reduced support for those who survived' (Robertson 2004: 6). Such assumptions of responsibility are made all too easy by the prevalence of victim-blaming strategies that have emerged around, and continue to cloud, considerations of occupational deaths and injuries.

Thus the term 'accident' both invokes and reinforces resort to an individualising lens for viewing and 'understanding' safety incidents, and this lens extends to quasi-legal approaches to such events. Law's complex reasonings in relation to defining crime, while not exclusively focused on the individual, have an individualising effect which extends beyond the notion of intent *per se*. Thus, even where intent is not the issue in determining legal liability – such as in the case of corporate manslaughter – then the individualising ethos of criminal law has militated against such successful prosecutions, whilst in contexts where this charge has been raised, such as following the Zeebrugge or Southall 'disasters', then charges of manslaughter have been raised against relatively low-level individuals on the scene of the incident – namely, the assistant bosun or the train driver (see Tombs 1995, Slapper and Tombs 1999: 30–4, 101–107, and Chapter 10, *passim*).

These legal responses are reinforced by, and themselves reinforce, the immediate 'commonsense' appeal of accident-proneness. This is clearly illustrated in media coverage of the recent disasters in the UK. In the immediate aftermath of many of these incidents, there emerged initial popular attributions of blame directed against individuals on the scene of these 'accidents'. In such cases, the power and appeal of accident proneness is only diminished through the long detail of a public inquiry, at which there is (often) a complex tracing of events surrounding an incident in order to arrive at its 'fundamental' cause(s), located within managerial and organisational systems.

Such inquiries may be contrasted with the general lack of a search for 'fundamental' causes following minor, 'day-to-day' incidents, causes thereby remaining obscured. Hughes has noted how 'social attitudes to safety tend to blame an individual for his own death through carelessness or bad luck', whereas following disasters there is at least more likelihood that there will be a public focus upon 'management' or 'systems' (Hughes 1984 : 15).

These types of argument, then, are espoused (and believed) not only by those individuals and groups that may have a vested interest in seeing their acceptance – in particular, here, we think of employers who, were workers not thus blamed, might find themselves being held responsible for a much larger proportion of industrial accidents with the possible consequence of coming to be considered as blameworthy – but indeed also by other groups who have, or have had, a direct interest in industrial health and safety – in particular, here, we refer to governments and their appointees, but also to health and safety 'experts' or professionals and, most surprisingly, to some workers themselves (Leopold 1985: 265; Watterson 1988: 293).

The practical nature and effects of the ideology of accident-proneness and worker-blaming are perhaps most clearly exemplified in the reactions of Union Carbide Corporation (UCC) to the gas release at Bhopal (see Chapter 1) – in this particular case, the 'incompetent' worker is invoked although, through the use of racist and imperialist stereotypes, this more specifically becomes the incompetent *Indian* worker (Everest 1986: 107). Thus in attempts to 'explain' the circumstances surrounding the accident, UCC refused to countenance any suggestion that the methylisocyanate leak may have resulted from fundamental inadequacies in the Bhopal plant's basic design, state of maintenance, staffing levels, and so on. In fact, UCC claimed, variously, that the leak had been caused by: a failure on the part of operators to comply with basic safety procedures; a purely mistaken introduction of water into an MIC storage tank by an unwitting employee; an act of terrorism by Sikh extremists; and, most recently, and it appears 'definitively', an act of sabotage by a disgruntled employee. More generally, UCC cast the Indian workers (and, indeed, government regulators) as technologically unsophisticated, and incapable either of understanding or adhering to basic safety procedures (ibid.).

It is not the intention here to enter into a discussion of the causes of the Bhopal disaster, nor to engage in a detailed critique of UCC's final version of the 'sabotage theory'.[7] It is sufficient to state here that, although it took UCC nearly three-and-a-half years to develop this 'theory', the idea of sabotage as the main cause of the accident had been mooted by UCC in the immediate aftermath of the tragedy; thus there is a real sense in which UCC *assumed* sabotage to be the cause of the disaster and then set out to prove in detail how this could have been the case (see, for example, Kalelkar 1988: 557, emphases added).

UCC is not exceptional. But the Bhopal disaster is particularly interesting for us, because the almost automatic responses of UCC to the 'accident', plus the difficulties encountered subsequently by those attempting to oppose such 'common-sensical' accounts of its causes, attest to the significance – that is, the deep-rootedness and pervasiveness – of the concept of 'accident-proneness'; and this is so even where the incident has been subject to as much publicity, investigation, and even court hearings, as has this one.

In general, then, the most important consequences of the ideological thrust of 'accident-proneness' is to assign responsibility for, and identify the objects of improvement in relation to, industrial injuries. Accepting the nature of the workplace and work processes as given means that the management of safety becomes a task of controlling the potentially hazardous actions or omissions of workers. Further, the predominance of the notion of accident-proneness meant that only *certain types* of experts entered the field of safety and health (those concerned with health sciences and the law, rather than, for example, engineers or designers of work organisation); moreover, this ideology remained virtually unquestioned by these experts, so that the very expertise which they possessed, which might in fact have proved to be a resource upon which workers could draw in the course of struggles over health and safety, was very often exercised upon workers, in the sense of being used to justify an ever-increasing regulation of the details of their behaviour, both inside and outside the workplace (Lessin and O'Neill 2002).[8] The issue becomes behaviour modification through control of workers rather than modification of dangerous workplaces – and, coupled with reward programmes for 'accident-free' working, through creating a culture of non-reporting of industrial injuries, such programmes have in-built tendencies to appear 'successful' (ibid.).

At a 'macro' level, then, the acceptance that most 'accidents' are caused by workers has key implications in terms of the necessary level and nature of regulation: that companies do not necessarily need to be subjected to more stringent state regulation in the sphere of health and safety, but rather should be left, to a large degree, to self-regulation. As Harrison has put it in her historical analysis of the role of the idea of an accident in factory regulation, it is 'a means of dealing with the difficulty for factory administration of directly interfering with employers' interests' (Harrison 1993: 269). Such approaches within both the US and the UK, then, need to be linked to the increasingly strident logic of self-regulation being espoused in many parts of the advanced capitalist world. In short, victim blaming and the 'concept' of accident-proneness are analogous to other key elements of capitalist ideology – they individualise structural issues. At a 'micro' level, then, companies, being left to self-regulation, and defining the problems of health and safety in terms of their workforce, can engage in a stricter disciplining of, and control over, the behaviour of that workforce, so that individual workers find themselves subjected to increasing levels of regulation over the minutiae of everyday behaviour in the workplace (and, indeed, outside of the workplace, too). The recent re-emergence of behaviour-based safety programmes (Lessin and O'Neill 2002) – essentially forms of behaviour modification which cover a range of techniques from disciplining to training – also involve a form of *responsibilisation* of individual workers. As recent criminology has demonstrated in the context of crime control, responsibilisation of individuals is integral to a realignment of the state's responsibilities.

The possible pincer-like combination of these macro- and micro-effects is stunningly exemplified in new regulatory practices which impose tickets (akin to parking tickets) for health and safety violations. Discussing this new practice in Ontario, Gray has argued that under 'neo-liberal self-regulation, workers are essentially forced to police their own hazards through individual responsibility', and that such an approach

> is a key example of the diffusion of employer responsibility and blameworthiness for unsafe working conditions. The ticketing practice not only directly polices workers in health and safety, but it also forces workers to practice individual responsibility. Workers must not only enforce regulation themselves to avoid unsafe work (a physical harm), but they must also be cautious of health and safety tickets (a financial harm) (Gray 2006: 879).

Indeed, a study of the rationales for the types of violations to which tickets might be attached found that these were all of a 'clear and visible' nature, and that they excluded both 'organisational factors' and 'organisational/managerial decision-making' (Gray, 2006). This, then, represents a clear blurring of responsibility for unsafe conditions, as well as 'the notion of who is a health and safety offender' (Gray 2006: 888). And its very existence is, if not wholly explained by, logically related to, ideas of accident-proneness and associated tactics of victim blaming.

Safety crimes as violence?

Now, if safety crimes tend to be obscured, rendered (relatively) invisible, while the phenomena to which this label points are in fact misrepresented as 'accidents', then one issue raised is the extent to which academic social science has sought either to address this misrepresentation, or to bring to light a relatively hidden social problem – each, surely, key rationales for academic investigation. Since our particular concern here is with the excavation of crimes, then the obvious discipline to which we might turn is criminology, in order to assess how safety crimes are considered therein. Now, on one level, this query can be answered very curtly – for if criminology has historically marginalised the crimes of the powerful, then it has paid particularly scant attention to safety crimes. For us, while this is hardly surprising, it is in one respect very curious – for within criminology there is a significant level of attention devoted to the phenomenon of violence in general. Yet if it is within this rubric of 'violent crimes' that we would argue safety crimes should be considered, any survey of criminological work on violence would conclude that such considerations are notable for their absence.

In short,[9] while there are obviously enormous epistemological, theoretical and political differences between the various approaches to violence taken within what stands for academic criminology, these approaches share, virtually without exception, certain characteristics or, more accurately, assumptions – assumptions which essentially rule out any consideration of safety crimes. Thus crucial within the definitions of violence deployed or included within these approaches are two central assumptions: first, a primacy granted to intention; and, second, a focus upon individual as opposed to collective sources of violence, and thus the centrality of violence as inter-personal as opposed to structural.

Now, intent enjoys significant legal status – while, relatedly, for some it is a key distinguishing criterion between conventional ('real') crime and corporate crime (merely technical offences) (see Pearce and Tombs 1998: 231). However, it is worth noting here that the notion of intent presupposes, and then concretises, a moral hierarchy which, once examined, is counter common-sensical. This point was first made – to the best of our knowledge – by Reiman in a simple but striking fashion. Reiman contrasts the motives (and moral culpability) of most acts recognised as intentional murder with what he calls the indirect harms on the part of absentee killers, by which he means, for example, deaths which result where employers refuse to invest in safe plant or working methods, where manufacturers falsify safety data for new products, where illegal discharges are made of toxic substances into our environment, and so on. Reiman notes that intentional murderers commit acts which are focused upon one (or, rarely, more than one) specific individual, a point which we know holds for contemporary Britain, despite moral panics about 'stranger danger'[10]; thus in such cases the perpetrator – whom in many respects fits our archetypal portrait of a criminal – 'does not show general disdain for the lives of her fellows' (Reiman 1998: 67). Reiman contrasts such forms of intentional killing with the deaths that result from 'indirect' harms. For Reiman, the relative moral culpability of the intentional killer and the mine executive who cuts safety corners is quite distinct and, he argues, contrary to (indeed diametrically opposite to) that around which criminal law operates. Thus the mine executive

> wanted to harm no-one in particular, but he *knew his acts were likely to harm someone* – and once someone is harmed, the victim is some-

> one in particular. . . . there is no moral basis for treating *one-on-one harm* as criminal and *indirect harm* as merely regulatory. (Reiman 1998: 67–70, original emphases)

Thus, Reiman concludes, offenders of intentional, one-on-one harm are less likely to represent some generalised threat to others than the mine executive. The reasoning is convincing, and points to indifference or 'indirect' harm as at least, if not more, culpable than intention and 'direct' harms – with implications for how these are treated by any criminal justice system (see Pemberton 2004). Yet the greater moral culpability that is attached both legally and popularly to acts of intention can also allow those implicated in corporate crimes to rationalise away the consequences of their actions – techniques of neutralisation made possible through, and supported by, key institutions such as the media, formal political debate, and so on (see above, also Slapper and Tombs 1999: 105–7, 118–22; and Chapter 5 *passim*).

If intent is central to dominant legal and academic understandings of violence, also significant is the primacy attached to explanations at the level of the individual. As Salmi has written, the 'usual treatment of violence' is infected with '[e]xcessive individualisation', 'attributing solely to individual factors actions that cannot in reality be accounted for in individual terms. By so doing, the possibility of a causal link between the violence observed and the surrounding social structure is systematically dismissed' (Salmi 1993: 8). Rather, focus remains upon 'the individual and the eradication of such deplorable behaviour' (Catley 2003: 4). This is unsurprising in advanced capitalist societies, since analysis at this level coheres entirely with the ethos of individualism upon which such societies are maintained.

Certainly once one abandons an epistemological commitment to individualism, then more encompassing definitions and considerations of violence become possible. For example, among the work on the issue of workplace violence, that of Bowie stands out, since he has sought to develop the category of 'organisational violence', and although this is partially limited by some commitment to intention, it is far more useful than other definitions of violence. Thus, for Bowie, organisational violence 'involves organisations knowingly placing their workers or clients in dangerous or violent situations or allowing a climate of abuse, bullying or harassment to thrive in the workplace' (Bowie 2002: 6). If this still retains some commitment to intention, it at least moves beyond simple understandings of individual action, not least because it acknowledges how a general organisational demeanour of generating or turning a blind eye towards violence can be fostered 'in a growing economic rationalist climate of decreasing job security, massive retrenchments and expanding unemployment that pitted workers and unions against employers' (ibid.: 9). Of particular interest for us is that Bowie also notes that such violence is much harder to recognise due to the tendency to 'blame' (ibid.: 6) individuals, and to develop strategies for responding to violence which are 'based on a pathology model of "mad, bad or sad" employees or clients and patients who are seen as individually responsible for the violence occurring at work' (ibid.: 8).

Similarly, Hills, in introducing a collection of case histories of 'corporate violence', defines this phenomenon as

> Actual harm and risk of harm inflicted on consumers, workers, and the general public as a result of decisions by corporate executives or managers, from corporate negligence, the quest for profits at any cost, and wilful violations of health, safety and environmental laws (Hills 1987: vii).

Through the cases presented in this collection, and in his final considerations regarding these, Hills concludes that such violence is understood 'not in the pathology of evil individuals but in the culture and structure of large-scale bureaucratic organisations within a particular political economy' (Hills 1987a: 190). Hills's understanding of violence seems to shift beyond both intention and individuals. These shifts beyond intention can also be discerned in some other, recent criminological work which, if not explicitly couched in the language of violence could be reframed as such – we are thinking in particular here of a variety of work within the rubric of 'green' criminology (Lynch and Stretesky 2003; White 2003), within which we would include the systematic (state-corporate) exploitation of whole classes of people (Walters 2006). Indeed, across such work, the use of 'violence' starts to resemble some of the characterisations of the Holocaust, of genocide, and of other forms of state violence, and even in this very fact, the marginality of such modes of arguments to, and substantive focus within, criminology becomes apparent. So if the (partial) transgressing of the boundaries of dominant understandings of violence by Bowie and Hills are not unique, they are, within criminology, extremely rare. All too often, one must move beyond criminology – as we did in Chapter 1, where we utilised briefly the work of Salmi – to arrive at more accurate ways of describing, analysing, then theorising about violence.

Conclusion

We have sought to demonstrate in this chapter how safety crimes are systematically obscured through a series of mutually reinforcing processes. At the broadest level, safety crimes remain socially and politically invisible in ways which mirror the invisibility of corporate crimes in general. Thus there is an array of social processes that contribute to removing such offences from dominant definitions of 'crime, law and order' (Slapper and Tombs 1999).

The obscuring of the criminal aspect of safety crimes has been crucially generated and strengthened historically and contemporaneously by the idea of the accident-prone worker, which in turn legitimates strategies of victim blaming, that is, assigning responsibility for, and identifying the objects of improvement in relation to, industrial injuries onto individuals rather than the nature of the workplace, work processes, managements, cultures of an industry or sector, and so on – that is, the very causal factors we highlighted as crucial in producing safety crimes in our presentation of case studies in Chapter 1. The ideology and practices of accident-proneness thus need dismantling if we are to treat 'accidents' as (potential) safety crimes.

Indeed, once occupational injuries are viewed not as accidents but as incidents which are not only largely preventable, but which the law requires to be prevented, then they fall within the ambit of criminology. Then, as we have argued in this text, if we consider these illegalities in terms of their consequences – injury and death – we realise that these look remarkably similar to the results of those events that are deemed as 'proper' violence. However, as we have seen in the latter section of this chapter, dominant criminological conceptions of violence are based upon an implicit but tenaciously-held association of violence with both the inter-personal and with intention, both qualities that are often absent from safety crimes. That safety crimes are crimes of (actual or potential) violence is, for us, unequivocal; yet to reach this conclusion one must generally move beyond criminology, to some well-accepted understandings of violence which allow us to move beyond the interpersonal and intention, and within which safety crimes clearly fall.

It is the very organisational locus or origins of safety crime which makes the emphasis upon either an illegal *act* or an *intention* usually inappropriate, for two reasons. First, each term is anthropomorphic and individualising, and by definition is problematic in its application to a corporate entity which is more than, or different to, the sum of individual human actors. Second, intention in particular implies a relatively unproblematic link between an act or omission and its consequence, yet this simplistic causal sequencing leads to an obscuring of the construction and maintenance of a situation or context which, as a consequence, is fertile ground for violations. We reiterate a point we have made several times in this text: any focus upon safety crimes requires us to examine these in terms of their *organisational production.*

In Chapter 2, we sought to set out the extent of occupational deaths and injuries across British workplaces; we made clear at that point, however, that we were speaking of incidents rather than crimes *per se* – that is, not all deaths and injuries are the result of safety crimes. In this chapter, we have further considered the nature of these phenomena, and paved the way for considering some of these deaths and injuries as safety crimes, indeed crimes of violence. In the following chapter, Chapter 4, we turn directly, through a combination of theoretical and empirical arguments, to an attempt to reveal the extent of corporate violence in the form of safety crimes.

Notes

1 And had not Blunkett resigned following the nanny/passport/Kimberley Clarke row, he may have continued in the office, thus lending even greater continuity to this Office.

2 Safety is not of course unique in some of these respects; prisons, for example, have been presided over by a long line of junior ministers since 1997. However, the prisons minister *is* located within the Home Office. Moreover, as we argue throughout this section, the key issue is not about any specific element of the processes we describe, but the fact that they are all mutually reinforcing.

3 Such figures are difficult to establish beyond doubt, of course. This claim began to circulate on the basis of a calculation made by the Liberal Democrats Press Release at www.libdems.org.uk/story.html?id=8180 and also BBC NEWS/POLITICS, May 12, 2005, at http://newsimg.bbc.co.uk/1/low/uk_politics/4541377.stm. Given the source of this claim it needs to be treated with particular caution – but it remains that there is no doubt that a great deal of new criminal offences have been created, whether or not these can actually be counted in four figures.

4 In fact, it was only in 2005 that it was finally decided by the CPS that no individuals would face charges over the 1999 Ladbroke Grove rail crash which claimed 31 lives, the CPS citing 'insufficient evidence': Network Rail, formerly Railtrack, did face criminal charges under the Health and Safety at Work Act 1974 (BBC News, December 6, 2005, http://news.bbc.co.uk/1/hi/uk/4504696.stm).

5 Although without being named formally, claims which amount to accident proneness were in circulation well before then; see, for example, Harrison 1993, on factory inspection at the turn of the nineteenth century.

6 And thus not deserving of the state's support. Contrast with New Labour's (claimed) championing of the victim as being at the centre of the criminal justice system.

7 The former has been dealt with exhaustively, most adequately in Everest 1986; Jones 1988; Morehouse and Subramaniam 1986; and Shrivastava 1992, while the latter has been attempted in Bergman 1988 and Bhopal Action Group 1988. See, more generally, Pearce and Tombs 1998: 194–219.

8 In the previous chapter, we referred to recent research findings that about 1,000 deaths on the roads are actually work-related fatalities, concluding that 'Britains roads are the country's most dangerous workplace as under-pressure workers, struggling to meet deadlines and suffering fatigue from long hours, become a danger to themselves and others' (Trade Union Congress 2005b). In response to this, a Cranfield School of Management expert on driver behaviour said: 'Companies need to conduct psychological profiling of employees to see who has a tendency towards aggression or thrill seeking' (Campbell 2005).

9 A detailed examination of the taken-for-granted assumptions within criminology regarding what constitutes violent crimes – and the implications these assumptions have for any potential criminological focus upon safety crimes – can be found in Tombs, 2006.

10 BCS figures for 2002/03 show that 'In over half of violent incidents the offender/s were known to the victim in some way; in one-third of incidents they were known well' (Smith and Allen 2004: 11).

Chapter 4

Discovering safety 'crimes'

Introduction

In Chapter 2, we explored the extent to which we can treat official statistics on deaths and injuries at work as accurate, and argued that, whatever the limitations of the data, it is clear that occupational injuries constitute a major social problem. We have also argued that injuries and deaths at work are generally misrepresented when cast as 'accidents' – a label that has a series of problematic specific connotations and social effects. But our considerations thus far have not established the extent to which, occupational deaths and injuries can be regarded as crimes.

Now, it is important to be clear that in the context of a treatment of safety crimes, we cannot assume that all injuries are the result of some violations of the law. This assumption is made even in two of the best pieces of work within criminology that raise the issue of safety crimes: Box's (1983) classic chapter on 'Corporate Crime' in *Power, Crime and Mystification*, and Reiman's (1979) similar, US-based discussion in *The Rich Get Richer and the Poor Get Prison*. These classic texts of critical criminology each contain important, and at the time of publication largely novel, insights into the problem of safety crimes – yet they can also be criticised for treating in their data all workplace injuries and death as crimes. That is, in their attempts to criticise dominant understandings of the crime problem, they seek to include within the category of crime phenomena such as occupational injuries and ill health. Yet, while we certainly agree with the need to reformulate dominant constructions of 'crime', there is a danger in using the term 'crime' *too* loosely, without grounding it in the real processes of law enforcement and criminalisation. The danger, which these authors run into, of allowing subjective opinions to lead our analysis (and thus exposing our work to the charges of 'moral entrepreneurship': Shapiro 1983). A generally applicable definition of crime cannot be anything that we merely feel is wrong or distasteful. 'Crime', if it is to retain any meaning at all, must always be related to the material processes and institutions of criminal justice. To put it bluntly, just because someone is killed or injured at work does not necessarily mean death or injury is the result of a crime – this needs to explored then demonstrated, rather than assumed or asserted.

Our previous comments notwithstanding, a key message from the work of Box and Reiman is that the very notion of crime is contested. An important way of approaching crime as a contested concept has been developed by European writers who come from the 'abolitionist' perspective (Bianchi and van Swaaningen 1990; de Haan 1990). Abolitionists such as Hulsman (1986) point out that 'crime' has no intrinsic, objective property: it has no ontological reality outside the social institutions and processes that define it. What we know as crime should be understood as merely the types of behaviour and conduct that the state defines as worthy of control and punishment: some forms of harmful conduct are deemed worthy of control and others are not. This is a point that we will return to in the conclusion where we will argue that – counter to the received wisdom in criminology – the criminalisation of safety offences is entirely consistent with this perspective. The point to bear in mind at the moment is that crime is a relative concept which is defined in relation to the process of making and enforcing the criminal law. It is this point of critique that provides a starting point for understanding both why most injuries and deaths at work are not criminalised *and* how criminal law and criminal justice institutions perpetuate our understanding of them as 'accidents' or 'technical offences', and therefore not really crime at all.

Understanding the process of criminalisation requires us to broaden our horizon of inquiry beyond a narrow consideration of state-defined crime. The forms of conduct that are criminalised in society can be considered as part of a wider spectrum of social harms (Hillyard *et al.* 2004). It is particularly important when we are dealing with corporate harm to consider that conduct which is defined as crime will only ever represent the tip of the iceberg (Whyte 2004).

The intention of this chapter, however, is to show that the marginalisation of safety crimes within mainstream criminal justice discourse and practice cannot be justified with reference to the definition of crime set by the criminal justice system itself. At the same time, we would argue that to engage on the terrain of crime and criminal justice, we need to retain some workable definition of 'crime'. Describing all injuries and deaths at work as crimes may have some justification, but it reduces our approach to crime to a labelling or a moralising exercise.

We do not seek to resolve the contradictions that underpin the notion of crime here (although we will say something more about those contradictions in Chapter 6). The important thing for the purposes of this book is to recognise that 'crime' is a term that has no settled meaning, but that it always carries an ascribed, or socially constructed meaning. We recognise that the harms that result in criminalisation are not fixed in time, but are subject to prolonged struggles over their definition and enforcement. We also recognise that the process of criminalisation is not fixed either. Thus, as we shall see in Chapters 6 and 7, we can find at different moments in history movements both to criminalise and to decriminalise safety crimes.

Using official crime figures?

Despite their well-documented limitations, 'official crime statistics' remain predominant as a 'descriptive medium' (Maguire 1997: 139) in crime debates, constituting at least a starting point for almost all efforts to quantify the extent of various forms of offences. The particular concern of this chapter is with the availability – or otherwise – of official data on safety crimes, that is, infractions of a legal duty placed upon an employer or a corporate entity, this legal duty defined within the framework of the criminal law, most notably the Health and Safety at Work Act (HASAW Act) 1974.

Even a cursory examination of 'notifiable offences' for England and Wales reveals the extent to which their focus is almost exclusively upon 'conventional' crimes. Home Office data collated under the category 'Violence against the person' includes the following categories and subcategories of violent offences:[1]

- *Homicide*, including: murder, premeditated and unlawful killing of another person; manslaughter, unintentional killing of another person; infanticide, intentional killing of an infant under 1-year-old by a mother suffering from post-natal depression or other post-natal disturbance.
- *Causing death by dangerous driving*: killing another person by driving a motor vehicle dangerously on a road or in a public place.
- *Causing death by careless driving when under the influence of drink or drugs*: killing another person by inattentive driving, the driver having recently consumed alcohol or drugs.
- *Causing death by aggravated vehicle taking*: killing another person by dangerous or careless driving of a stolen motor vehicle on a road or in a public place.
- *More serious wounding or other act endangering life*: viciously intending to cause grievous bodily harm to another person.

- *Other more serious violence offences*, including: attempted murder, attempting to kill another person in a premeditated and unlawful manner; threat or conspiracy to murder; stating an intent to kill or solicit, encourage, endeavour, or persuade someone to do so; child destruction, intentional killing of an unborn child capable of being born alive by its mother; endangering railway passengers, placing railway passengers in danger by interfering in any way with the railway system.
- *Less serious wounding*, including: maliciously inflicting grievous bodily harm, with or without a weapon' and 'assaulting someone and causing him or her actual bodily harm'; racially or religiously aggravated other wounding, 'maliciously inflicting grievous bodily harm, with or without a weapon' and 'assaulting someone and causing him or her actual bodily harm, where there is a racial or religious motive to the offence'; common assault (includes some minor injury); common assault (no injury), 'assaulting another person where the victim receives a minor injury or … no injury'; racially or religiously aggravated common assault, 'assaulting another person where the victim receives no injury and there is a racial or religious motive to the offence'; harassment, 'putting people in fear of violence; also continual, persistent attacks causing alarm or distress'; racially or religiously aggravated harassment, 'putting people in fear of violence; also continual, persistent attacks causing alarm or distress where there is a racial or religious motive to the offence'; possession of weapons, 'being found in possession of an object or instrument', and assault on a constable, 'when a police officer is assaulted in the course of his or her duty'.

Now, *none* of the above categories – as diverse and all-encompassing as they appear to be on first reading – include occupational injuries. One obvious category for including records of safety crimes is that covering homicide, manslaughter and violent crime. However, the crime of manslaughter in England and Wales has been legally and socially constructed in a way that renders it inapplicable to corporate offences/offenders, not least because of the centrality of *mens rea* ('knowing mind') to this offence (see Chapter 7). Perhaps a more fruitful source of data on safety crimes is the offence category of manslaughter (culpable homicide in Scotland). The problems in attempting to apply this charge successfully to both the corporate entity and/or individual employers are well-known (see Slapper and Tombs 1999; Tombs and Whyte 2003a), and given that, at the time of this book going to press, at the end of January 2007, just fourteen successful manslaughter prosecutions for safety offences have been recorded,[2] such data are unlikely to make much of an impact in the relevant Home Office data column. And the sub-categories of wounding and assault might also legitimately include reference to infractions of the criminal law which resulted in injury (fatal or otherwise) to an employee or member of the public. Again, this is not actually the case. One of the most striking aspects of legal – not to mention political and academic – treatments of violence is the general absence of the production of violence by corporate activities (Wells 1995; Pearce and Tombs 1992; Tombs 2006), an issue upon which we focus in Chapter 3.

Despite the problems with working with the available death and injury data (see Chapter 2), the complete absence of safety crimes from mainstream crime figures urges us to draw comparisons, no matter how crude, between homicides recorded in the Home Office notifiable offences and deaths and injuries at work. First, in terms of deaths, it is possible to compare the numbers of people killed at work with those recorded by the Home Office as homicides (that is, murder, manslaughter and infanticide).

Now, from these data we can make several observations. Initially it appears that one is more than twice as likely to be a victim of homicide in England and Wales than to die as a result of an acute workplace-related incident. However, against this we need to bear in mind that the fatality data above is for workers only – that is, it excludes members of the public. Further, as we indicated in Chapter 2, the headline figure that Table 11 uses only captures somewhere between one-seventh to one-eighth of the occupational fatalities which we know about for the most recent year, 2004/05. On the basis of these qualifications, being a victim of a work-related fatality looks much more likely than being a victim of homicide.

Table 11 Reported fatal injuries to all workers and rates per million workers, 1996–2005[3]

1996/ 1997	1997/ 1998	1998/ 1999	1999/ 2000	2000/ 2001	2001/ 2002	2002/ 2003	2003/ 2004	2004/ 2005
287	274	253	220	292	251	227	236	223
11	10	9	8	10	9	8	8	8

Table 12 Homicides in England and Wales, 1995–2005 and rates of homicides per million population, England and Wales, 1995–2003[4]

1996	1997	1997/ 1998	1998/ 1999	1999/ 2000	2000/ 2001	2001/ 2002	2002/ 2003[5]
678	735	731	744	763	849	863	1043
11.4	11.0	11.9	12.6	13.2	15.1	15.7	19.3

A similarly crude, but instructive, comparison can be made in terms of injuries. According to the 2003/04 British Crime Survey, there were a total of 2,780,000 violent offences in England and Wales, and 4.1 per cent of people experienced a violent incident, half of which resulted in no injury.[6] Smith and Allen (2004: 1) noted that 'The majority of violent crimes involve no significant injury to the victim and about half (51 per cent) involve no injury at all. In the 2002/03 BCS, 11 per cent resulted in medical attention from a doctor and 2 per cent in a hospital stay'. In terms of risk of violence, respondents to the 2002/03 BCS were calculated to be at a 4.1 per cent – or 41 in a thousand – risk of being a victim of violent crime (ibid.: 6).

If we turn back to Tables 2 and 3 in Chapter 2, we can indicate some illuminating comparisons with violent crime recorded by the BCS. In terms of the injury data, we need to recall the high levels of under-reporting, particularly amongst self-employed workers and particularly for minor injuries, but even if we take the over three-day injury rate for all workers (that is, employees and the self-employed), we find a rate of 415.25 per 100,000 for the most recent year (2004/05). Putting this in percentage terms, this means that 4.15 per cent of workers have reported an injury that caused them to be absent from work for three days or more – this compared to 4.1 per cent of BCS respondents citing their victimisation to violent crime. What is important here is to recall that over half of the latter incidences of violence resulted in no injury at all, and only 11 per cent resulted in any form of medical attention.[7] Then, if we turn to the data in Table 2, in Chapter 2, we find the percentage of workers experiencing a major injury – defined by HSE as 'fractures (except to fingers, thumbs or toes), amputations, dislocations (of shoulder, hip, knee, spine) and other injuries leading to resuscitation or 24-hour admittance to hospital' – standing at 1 per cent (100.3 in 100,000). This can be compared with the *less than 0.4 per cent* of BCS respondents (11 per cent of the 4.1 per cent who experienced violence) needing medical attention. The comparisons are crude, unsatisfactory, beset by methodological and statistical problems, so that the exact figures are almost meaningless – but they are broadly indicative of a rather undeniable conclusion, namely that work is more likely to be a source of violence in Britain than those 'real' crimes recorded by the Home Office.

The absence of safety crime data from Home Office statistics outlined above is significant in terms of the separate ontological status accorded to safety crime. The fact that the safety crime count is not published by police forces or by the government department responsible for crime, but by a separate agency, reinforces the exclusion of safety crime from mainstream debates on crime. Constructing safety crime as something that has to be acted on and counted not by police forces, nor by the Home Office or Scottish Executive, but by regulatory agencies, reinforces the idea that safety crime is not 'real crime'. The institutional segregation of safety crimes by the state therefore has profound implications for how we think about safety crimes. Further, this means that we must look to what those agencies do if we are to discover more about how safety crime is actually defined: it is in investigative, administrative, enforcement and juridical processes that crime is recognised and defined (Alvesalo 2003a).

Normally in criminal justice systems we can find a distinction made between *regulatory offences* and *crimes of violence*. While it is important to recognise that regulatory offences are a sub-category of criminal offences, it is also important to recognise the features that distinguish them from other (mainstream) criminal offences. There are three features of the regulatory process that act upon the definition of safety crimes. First, the investigation and prosecution of safety crimes is in the remit of *regulatory* authorities rather than police forces. This procedural feature of safety crimes differentiates those offences immediately from the mainstream criminal justice system. This has an important bearing upon how safety crimes are regarded by their perpetrators, victims, and the wider public. If the police are not involved in the investigation of safety crimes, then those crimes are less likely to be socially constructed as 'real' crimes. It is their enforcement by a specialist organisation, the HSE, that is the most important factor in distinguishing regulatory offences from other criminal offences. In England and Wales, the relatively rare investigation and prosecution of safety crimes as manslaughter (as opposed to Health and Safety at Work Act) cases involve the police in collaboration with the regulatory authorities. Second, this process of administrative differentiation also places responsibility for a huge number of offences in the hands of a relatively small agency. The average workplace can expect a visit approximately every 20 years (Centre for Corporate Accountability 2004: para 3.30). This limits the extent to which safety crimes can be identified and processed, since offences will generally only come to the attention of HSE inspectors during investigations, routine inspections and 'blitzes' (discussed below). The third feature that allows us to distinguish between regulatory crimes and crimes of violence is the way that the criminal law has developed specific procedures for dealing with safety crimes. Regulatory offences often involve the use of low level pre-court penalties or administrative notices which can be issued by inspectors on the spot (normally improvement notices and prohibition notices). This automatically creates a question about whether we can described such offences, dealt with outside the normal rules of criminal procedure, as 'crimes'. Closely related to the previous point is that regulators very often seek to deal with many of the offences they come across informally. We pick this point up later in the chapter in a discussion of compliance approaches to regulation.

The gap between 'enforceable' and 'enforced' crime

We set out our definition of safety crimes earlier as 'infractions of a legal duty placed upon an employer or a corporate entity, this legal duty defined within the framework of the criminal law.

Yet, the points made in the preceding section highlight the fact that, very often, the differential treatment of safety crimes – the informal way that safety crimes are often dealt with by regulatory agencies – produces a gap between the breaches of criminal law that are uncovered by the authorities on the one hand, and the breaches that are processed and recorded as such on the other. Indeed, it was this point that prompted Edwin Sutherland to argue for a broad definition of white-collar crime. It is worth exploring this argument in a little detail here.

In a series of papers, articles and a book published between 1940 and 1949[8] Sutherland developed the concept of 'white-collar crime' – 'a crime committed by a person of respectability and high social status in the course of his occupation' (Sutherland 1983: 7). He thus challenged the stereotypical view of the criminal as typically lower class since 'powerful business and professional men' also routinely commit crimes. Criminal acts are not restricted to those dealt with in criminal courts. Other agencies such as juvenile courts may deal with 'violations of the criminal law' and some offences can be dealt with by either criminal courts or civil courts. Some individual white collar offenders avoid criminal prosecution because of the class bias of the courts – although businessmen could often be charged as accessories to such crimes as bribery unlike politicians they usually escape prosecution – but more generally they are aided by 'the power of their class to influence the implementation and administration of the law'. Thus the crimes of the upper and lower classes 'differ principally in the implementation of the criminal laws that apply to them' (Sutherland 1940: 35–37). Given that 'upper class' criminals often operate undetected, that if detected they may not be prosecuted, and that if prosecuted they may not be convicted, Sutherland argued that the criminally convicted are far from the closest approximation to the population of violators.

In his 1945 article 'Is "White-Collar Crime" Crime?' Sutherland set out to produce a more encompassing and abstract definition of crime. Crime requires the 'legal description of an act as socially injurious and legal provision of a penalty for the act' (Sutherland 1945: 132). Notice that while Sutherland holds on to the law in his definition of crime, he extends relevant bodies of law beyond the criminal law. He recognised that many laws which are enforced by administrative bodies through the civil courts also regulate actions which cause injuries to specific individuals or which undermine social institutions, and they also routinely impose punitive sanctions. Moreover, *contra* the view that such acts are merely 'technical violations and involve no moral culpability' in fact they are 'distributed along a continuum in which the *mala in se* are at one extreme and the *mala prohibita* at the other' (ibid.: 139). The content of laws and such legal distinctions are themselves social products (Sutherland 1945).

In other words, Sutherland argues that crimes are illegalities which are contingently differentiated from other illegalities by virtue of the specific administrative procedures to which they are subject. The corollary of this is that 'successful' criminalisation of the illegalities of the powerful would pre-empt them from arguing that their illegalities are merely 'regulatory offences', merely *male prohibita*. And there are persuasive arguments that it *matters* both practically and ideologically whether something is defined as a crime or a civil offence: that is, to be subject to one rather than another set of procedures has important differential effects.[9]

Sutherland's argument regarding the most appropriate definition of crime – itself essentially an argument for incorporating corporate and white collar offences into the study of crime – is a detailed one, much of which must be read alongside the responses of his critics, not least the legalistic, and well-known, response from Paul Tappan (1947)[10]. There is not the space to fully rehearse this debate here. Most fundamentally, though, Sutherland raises the crucial issue of how the relative power of different social groups affects what became criminalised and when:

> Embezzlement is usually theft from an employer by an employee, and the employee is less capable of manipulating social and legal forces in his own interest than is the employer. As might have been expected, the laws regarding embezzlement were formulated long before laws for the protection of investors and consumers. (Sutherland 1940: 36)

Nor is the 'meaning' of law closed nor simply to be read off from the avowed intentions of legislators. Thus he notes how the Sherman anti-trust legislation, while 'enacted primarily because of fear of the corporations', was mostly used as the basis for criminal prosecutions against trades unions in its first thirty or so years (Sutherland 1983: 57).

Further, he demonstrated the contingent nature of the distinction between criminal and other offences by tracing the genealogy of the laws regulating competition (antitrust), false advertising, labour relations, and infringements of patents, copyrights, and trademarks. Each:

> … has a logical basis in the common law and is an adaptation of common law to modern social organisation. False advertising is related to common-law fraud, and infringement to larceny. The National Labour Relations Board Law, as an attempt to prevent coercion, is related to the common-law prohibition of restrictions on

> freedom in the form of assault, false imprisonment and extortion. For at least two centuries prior to the enactment of the modern anti-trust laws, the common law was moving against restraint of trade, monopoly and unfair competition. (Sutherland 1945: 133)

Thus there are good reasons for suspecting that the differential application of law, the development of different legal categories, and distinct enforcement *modus operandi* for 'street' and corporate offenders are not rooted in any intrinsic differences in the offences *per se*.

Finally, Sutherland is quite clear that the differential interpretation and enforcement of law against white-collar crime is partially based upon the fact that legislators, judges and administrators within the criminal justice system are either subject to the material and ideological influence of business-people, or share common ideological and/or cultural worldviews (Sutherland 1945: 137–8). Concomitantly, when what Sutherland calls the 'status' of business people declines, so they become more prone to the development of rigorous enforcement of laws designed to regulate their anti-social conduct (ibid.).

Sutherland therefore approached the problems of accounting for the under-enforcement of white-collar crime by emphasising what is 'enforceable' rather than what is actually enforced. By doing so he managed to retain a robust notion of 'crime' – and the location of the deaths and injuries that can be logically described as crime within this definition – which allowed him to legitimately ask questions about the absence of white-collar crime from mainstream crime control agendas. He is therefore deriving his definition of white-collar crime from the 'gap' been what could be enforced and what actually is enforced.

Enforcing the law

An alternative approach to Sutherland's emphasis upon the 'enforceable' would be to link definitions of crime much more closely to the criminal, or at least to a formal legal process. For example, a definition of safety crimes could read as follows: acts or omissions which are dealt with by a criminal justice agency (including regulatory agencies with enforcement powers) using an enforcement procedure specified in criminal law.

The use of the words 'enforcement procedure' is significant here because, as Figure 1 indicates, enforcement notices issued by the UK HSE are a much more frequent response to safety crimes than prosecution. This definition excludes breaches of the law that are discovered but are not processed by criminal justice agencies. The weakness of this approach to safety crimes, therefore, is that it depends upon the agency's or, at a micro-level, the individual front line inspector's, decision to take action or not against safety crimes. In a given year, if enforcement notices or prosecutions for safety crimes are lower than the previous year, this is not necessarily because fewer crimes have been committed, but is more likely to be due to changes in enforcement practices, priorities, resources, and so on. This certainly creates a definitional weakness, but the strength of this approach is that it allows us to remain more clearly within the definitional boundaries of the criminal justice system.

Enforcement data, then, merely represent an index of offences against health and safety legislation that have – in differing ways – been subject to processing by regulatory agencies; that is, unlike injury data, enforcement data needs no reconstruction before we can speak unproblematically of safety crimes (at least on the definition of safety crimes used here). We must, however, keep in mind the key caveat that such data may tell us less about the volume of crime uncovered and more about the enforcement practices and priorities of inspectorates. (On this as a more general problem within criminological research see Coleman and Moynihan 1996: 32–9).

As we note above, breaches of safety law are more likely to be met by the issuing of various notices rather than prosecution. This works out at a ratio of around nine administrative notices issued by all enforcing agencies (HSE and local authorities) to every prosecution, a pretty consistent measure over the past five years or so. The figure below represents recent trends in HSE enforcement activity.

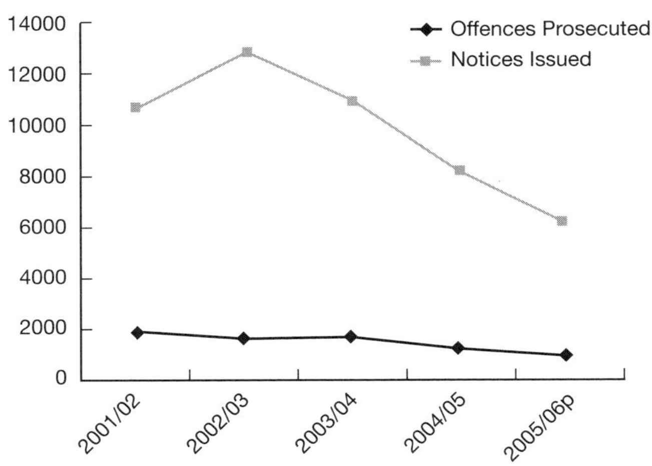

Figure 1 Cases prosecuted and notices issued by the UK Health and Safety Executive[11]

The first thing to note from Figure 1 is the consistently large gap between enforcement notices and prosecutions. Notices tend to be issued for lesser, but hardly non-serious offences, as the following quotation, from the most recent HSE Offences and Penalties report makes clear:

> A proportionate approach is taken to any breaches, so in less serious cases, the inspector will explain how the dutyholder is not complying with the law and advise them how to put the problem right … *Where the breach of the law is more serious*, the inspector may serve a notice to the duty holder. (Health and Safety Executive 2005: 4–5, emphasis added)

Three types of notice may be issued. An *improvement* notice requires breaches (offences) to be remedied within a specified period of time; a *deferred prohibition* notice is an order that a certain piece of plant or type of work activity will be shut down unless a breach is rectified within a specified period; an *immediate prohibition* notice – the most serious enforcement order – effects an immediate closure of a worksite, part of that worksite, or particular work process until a company complies with legislation of which it is in breach.

On the basis of the above, then, let alone in terms of what we know qualitatively about how inspectors actually enforce the law (see Chapter 7) – that is, on the basis of a general reluctance to use any kind of formal enforcement action – it is clear, then, that the total number of notices issued in any one year does provide a minimal indication of the numbers of safety crimes for that period.

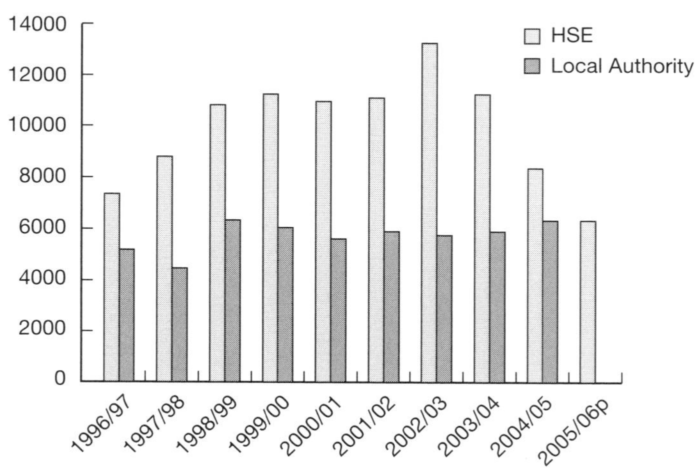

Figure 2 Total enforcement notices issued by UK authorities[12]

Now, again to be clear, these data tell us something about enforcement but, certainly on a year by year basis, are not indicative of the scale of safety offending. That is, they tell us less about the numbers of offences *encountered* by inspectors, much more about changes in enforcement policy, resources, year-on-year priorities, and so on.

Returning to Figure 1, the second thing to note from this graph is the recent steep decline in both enforcement notices issued, and prosecutions laid by, the HSE. We can thus observe a reduction of 42 per cent in enforcement notices issued over the most recent five years and an even steeper reduction of 49 per cent in prosecutions.

Prosecuting safety crime

What makes this decline even more striking is that it should be read within the context of historically low absolute numbers of prosecutions taken. This much is indicated by the data presented in Table 13 below, which compares prosecution data for nine years from 1975, roughly the period since the establishment of HSC/E[13] – with the nine years' data between 1996/97 and 2004/05.

Table 13 Prosecutions by inspectorates, 1975–1983 and 1996/97–2004/05[14]

Year	HSC/E	LA	Total	Year	HSE	LA	Total
1975	1588	78	1666	1996/97	1490	n/a	n/a
1976	1327	106	1433	1997/98	1627	506	2133
1977	1623	97	1720	1998/99	1759	424	2183
1978	1671	335	2006	1999/2000	2115	412	2527
1979	1373	211	1584	2000/01	1973	401	2374
1980	1443	307	1750	2001/02	1986	350	2336
1981	1260	291	1551	2002/03	1659	330	1989
1982	1427	296	1723	2003/04	1720	410	2130
1983	1366	261	1627	2004/05	1320	332	1652

It is clear that there is a slightly higher level of prosecutorial activity in the second than in the first period, notwithstanding the sharp drop in the most recent year for which confirmed data is available (2004/05). What is important for us here, however, is that we find across the years 1975–2005 both relatively consistent and absolutely low numbers of total prosecutions.

This rough consistency that Table 13 points to in the two periods is even more remarkable when one considers changes in a whole range of relevant variables across that 30-year period. These include, in no order of priority:

- dramatic changes in the macro-political climate, from the establishment of HSC/E during the high-point of corporatism, through the laissez-faire economics of Thatcherism, to the contemporary antipathy to any regulation of business on the part of New Labour;
- waves of both resource retrenchment and periods of increased resources;
- various shifts in enforcement priorities;
- significant changes in the economic structure, notably a shift from manufacturing to services, as well as a mushrooming of, and high turnover among, small businesses;
- a massive growth of self-employment;
- changes in economic performance, meaning that this 30-year period has witnessed several business cycles;
- changes in the organisation of production, summed up by many in terms of 'Post-Fordism';
- an increase in the absolute volume of law to be enforced, much of which has been EU-inspired.

If nothing else, the above data points to the enduring status of prosecution as the 'last resort' for regulators (Hawkins 2002). If there is a more recent decline, this is likely to be a result of the re-assertion of a concerted policy of de-regulation on the part of the Labour government that is beginning to undermine HSE enforcement strategy under intensifying budgetary pressures (and we will discuss this in more detail in Chapter 7).

The way that safety crimes are processed before they are considered for prosecution is crucial to understanding rates of prosecution. Regulators generally do not investigate or respond formally to all breaches of the law. Just as when a crime is reported to the police, at some point there will be a decision taken by the regulator about whether a reported incident is worth taking action over or not. As we have already seen in this chapter, most safety crimes in the UK, if they are acted upon in the first place, are dealt with using administrative notices. In other words, where inspectors opt for formal enforcement measures, these tend overwhelmingly to be procedures that fall short of prosecution. Indeed, in all categories, including the most serious injuries and deaths, only a minority of cases end up being prosecuted in the courts.

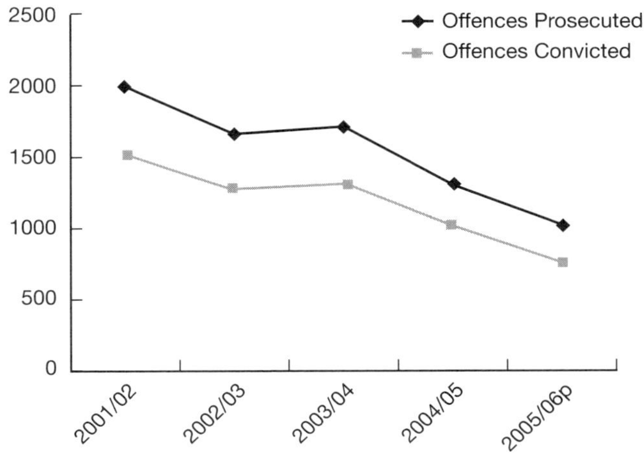

Figure 3 HSE prosecutions and convctions[15]

Figure 3 demonstrates a parallel decline in prosecutions taken by HSE and convictions for health and safety offences in the courts.

According to HSC enforcement policy, prosecutions are used where there is 'sufficient evidence to provide a realistic case of prosecution.' (ibid.: 12). Official policy statements such as this, however, mask the complexities of the prosecution process. Indeed, it is clear that the proportion of prosecutions to recorded injuries is very low indeed. To get an idea of this, we can set the total number of prosecutions, 652, against the level of recorded injuries for the same year, as set out in Tables 2–5, in Chapter 2. There we noted that for 2004/05, there were 593 fatalities to workers and members of the public, 31,702 major injuries to employees, 122,922 over three-day injuries to employees, and 14,316 non-fatal injuries to members of the public reported, even allowing for the various elements of under-reporting that were discussed. Thus, the ratio of total prosecutions to recorded injuries of varying severity (169,533 in total) is just in excess of 1:100.[16] If, as we argue in Chapter 2, that some two-thirds of injuries sustained in relation to work can be defined as 'crimes' then the ratio of prosecutions to recorded criminal injury appears very low indeed.[17]

Formal enforcement data therefore record only a tiny proportion of an as yet unquantifiable level of safety offences. The best that can be said regarding the extent of safety crimes on the basis of offence data is that the latter provides an absolutely minimal indication of the volume of the former. These data can therefore really only get us closer to understanding what the HSE do, as opposed to estimating the scale of safety crimes.

Low prosecution rates are to some extent driven by low detection. Thus, for example, some reported deaths are still not investigated, despite a HSE commitment to investigate all work-related fatalities. According to one analysis by the Centre for Corporate Accountability (CCA), in the five-year period 1996/97–2000/01, 75 worker deaths and 212 deaths to members of the public were not investigated (Unison/Centre for Corporate Accountability 2002). In cases of the most serious of non-fatal injuries, defined as 'major injuries', between 1996/7 and 2000/01, the percentage investigated almost doubled from 11 per cent to 19 per cent, with an increase in the actual number of major injuries investigated from 2,532 to 4,335; however, this increase still meant that, in 2000/01, *81 per cent of major injuries were not investigated.* Indeed, looking at the whole five-year period, some of the injuries to the most vulnerable workers remained un-investigated: there was no investigation into 905 of the 1,144 reported major injuries to trainees, or into 126 of the 164 injuries to those involved in work experience. Moreover, still by 2000/01, some of the most serious among the major injuries were not investigated, including: 72 'asphyxiations' (44 per cent of the total), 31 'electrical shocks' (35 per cent of the total), 333 'burns' (57 per cent of the total) and 418 'amputations' (41 per cent of the total; ibid.). In short, most safety crimes – including many of the most serious crimes – remain undetected and outside the radar of the criminal justice system.

This evidence also points clearly to a considerable variation in the chances of an incident being prosecuted across industry, region and type of incident. The research cited above found that 67 per cent of deaths of workers were prosecuted in 1998/99 compared with 11 per cent of major injuries. The large variation of prosecution across sectors is indicated by the finding that in the manufacturing and construction industry, 12 per cent of major injuries that were investigated were prosecuted. In agriculture, the comparable rate was 7 per cent and in the extraction industries (including the oil and gas industry), it was 3 per cent. The report found major discrepancies across regions. In Wales, for example, 20 per cent of major injuries were prosecuted, and in the West Midlands, the rate was 6 per cent. Although our concern here is with safety crimes rather than the much larger figure of occupational illness and ill health (see Chapter 1), it is worth noting that only 1 per cent of industrial diseases investigated by HSE were prosecuted.[18]

Those variations immediately alert us to the central role that the discretion of the inspector and the prosecutor has in those cases. In England and Wales the decision to prosecute (with the exception of corporate manslaughter cases where the decision rests with the Crown Prosecution

Service) is normally made by the HSE. In Scotland, the decision is made by the Procurator Fiscal after considering the evidence from an HSE investigation. HSE data set out in Table 14 below highlights starkly how this discretion can result in a variation of prosecution rates over time.

It is rarely the case that the failure to prosecute is merely down to insufficient evidence. It is the under-investigation of safety crimes and the lack of resources dedicated to inspection that are the key factors in keeping detection rates low. It is likely, then, that the very recent fall in rates of prosecution that we can see in Table 14 is directly related to a recent fall in inspection. As O'Neill (2006) has noted, the rate of inspection in the Field Operations Division of the HSE (the largest division in the inspectorate and one with comparatively high average rates of inspection) has reduced dramatically so that premises that could have expected on average one inspection every 6.99 years in 2001/02 could only expect one inspection per 10.84 years in 2004/05.

The data here indicates how a shift in prosecutorial decisions can produce dramatic changes – here, a decline – in absolute numbers and in the proportion of fatalities that result in a prosecution.

Table 14 Fatalities resulting in prosecution in the UK[19]

Year of fatality	Fatalities reported to HSE	Number of offenders prosecuted	% Fatalities resulting in prosecution
2000/01	350	149	43
2001/02	278	85	31
2002/03	256	68	27

At this point, it is important to recall our discussion in Chapter 2, where it was shown that the actual number of fatalities caused by working is seven to eight times the number of fatalities reported to HSE. Given that very few of the fatalities that are in this revised figure are likely to be subject to prosecution, the 'real', underlying rate of fatalities resulting in prosecution is likely to be much lower than the figures in Table 14 indicate.

Estimating safety crime

If this discussion indicates that, due to under-recording, under-investigation and under-enforcement, it is almost impossible to estimate the scale of offending in any complete sense, then an approach is to attempt to *estimate* the proportion of injuries and deaths (discussed in Chapter 2) that might be defined as 'crimes'. Given the acute problems with recording injuries also discussed also in Chapter 2, our most robust, albeit still hopelessly incomplete, measure of safety crimes is fatality data. At least we know pretty accurately which fatalities are included in the figures and which are excluded. Using this data as a baseline allows us to say *something* about the proportion of HSE recorded fatal injuries which might be the result of safety *crimes*.

First, there are three types of quasi-'official' evidence now available which allow some – albeit tentative – conclusions on the volume of this most egregious form of safety crime, namely: a series of special investigations into groups of fatalities undertaken by HSE in the 1980s; a now considerable stock of Commissions of Inquiry into a whole series of disasters that occurred in the UK; and a series of (HSE) Accident Investigation Reports, comprising one-off but highly detailed examinations of particular incidents. These have been discussed elsewhere (Tombs 1999b: 80–2), suffice to say here that each attempts to locate *responsibility* for injuries and incidents, and the focus of each is mainly on fatalities. The data presented in those various sources tends to the conclusion that in the clear majority of cases of workplace fatalities – in some two out of three fatal injuries – there is at least prima facie evidence of violations of duties placed upon employers by the HASAW Act, and thus at the very least a criminal case to answer. This general conclusion – based upon evidence indicating attribution of responsibility for the fatalities in question – also holds in the case of deaths at work whilst driving, with employers failing to meet their legal duties to provide safe systems of work, and to reduce risks 'so far as is reasonably practicable', where they are requiring employees to drive as part of their employment. Key causal factors in fatalities include: the failure to consider safer, alternative means of transport or indeed routes; the setting of unsafe schedules, journey times and distances; failures to maintain vehicles adequately; failures to invest in vehicles with additional safety features; and the lack of specialised training on offer for regular drivers (RoSPA 1998; see also Campbell 2005, Trades Union Congress 2005b).

To the above quasi-official evidence we might add a growing weight of academic research which attests to the fact that workplace deaths might usefully be explored as crimes, even in the absence of any legal process. While there has been relatively little attention to corporate crime in general within British criminology, the work that has been developed has often taken health and safety offences against employees as its focus.[20] Within this it is possible to discern work which provides an indication of the scale of such offending, though not in any complete quantifiable sense. Much of this work takes the form of case studies which re-create sociologically specific cases which were not dealt with adequately via the criminal justice system.

Thus work by Bergman (1991, 1993, 1994), Carson (1982), Pearce and Tombs (1993, 1997), Tombs (1995, 1996), Tombs and Whyte (1998), Slapper (1993), Woolfson *et al.* (1996), Woolfson and Beck (1997), and Whyte (1999a), for example, has all taken some aspect of safety crime as one focus, and all attest to the fact that the actual level of corporate offending not only far outweighs that recognised by official data, but is in some generalised sense comparable to conventional offending.

Exceptional within this tradition are the few attempts to re-analyse a series of deaths that have been processed (or not) through the criminal justice system. Notable here is the work of David Bergman and the work of Gary Slapper. Bergman (1991, 1994) re-analysed a subset of workplace deaths, both in terms of their causes and also by exploring the actual legal processes that did or did not follow each of them – with particular reference to the tendency of the 'near opaque filter' of the HSE 'preventing these cases from ever reaching the CPS's over-cautious scrutiny' (Bergman 1994: 2). Thus, his re-analysis of 28 deaths in the West Midlands between 1988 and 1992 concluded that:

> Four should have resulted in a manslaughter prosecution. Seven should have been referred to the police for a criminal investigation and subsequently to the Crown Prosecution Service. Eight were inadequately investigated by the enforcement authorities and require further investigation. Five should have resulted in a prosecution under health and safety law. (Bergman 1994: 90)

In only four of the 28 cases was the legal process found adequate.

The only comparable work to that of Bergman is Slapper's *Blood in the Bank*, which is based around a detailed analysis of 40 cases of death at work, following these through the inquest process.[21] For the record, juries returned verdicts of 'accidental death' in 33 of these cases and death by 'misadventure' in another five. Thus there was an essentially common finding in 38 of the 40 cases, or in 95 per cent of them – a figure considerably higher than that for verdicts in *all* inquests (that is, for all suspicious deaths, deaths in custody, and so on), where the equivalent figure is about 48 per cent (Slapper 2000: 98) – that deaths were caused by 'accidents' or 'misadventure'. Of the remaining two cases, one saw an open verdict returned, while in the other, death was found to be the result of 'the breakdown of safety procedures and lack of supervision'. The responsible company – British Gas – was fined £2,000 in this case, following a HSE prosecution under health and safety law (ibid.: 242–3). Such prosecutions – that is, for breaches of health and safety law – followed in 13 of the 40 cases examined in Slapper's book. All were successful and resulted in fines, ranging from £500 to £50,000. Ten of the 13 fines imposed were for £10,000 or under. Through his re-analysis of these deaths, Slapper argues that in 24 of the 40 cases, deaths were attributable to:

> the pressures of the profit driven economy. Thus, the work practice being used, the disregard for safety equipment or training, were the result not of ignorance but of unwillingness or inability to pay the necessary extra in order to ensure that the work would be safe. (ibid.: 164)

Such a conclusion indicates that there might be a far higher level of criminal culpability than is indicated by the fact that, to date, there have only ever been 14 successful prosecutions for corporate manslaughter following a workplace death in England and Wales, and no equivalent corporate homicide cases in Scotland.

The evidence documented in this section reinforces the importance of retaining Sutherland's distinction between what is enforceable and what is enforced. At the same time, the Slapper and Bergman studies discussed here attempt to understand the scale of safety crime whilst being guided carefully by a precise legal reasoning: the un-enforced crimes highlighted by those authors are only defined as crimes with close reference to the criminal law. So, even staying within the strict parameters of criminal legal reasoning, the conclusion is that most of the deaths analysed can be regarded as safety crimes.

Conclusion

In previous chapters we have demonstrated, first, that there are significant numbers of deaths and injuries caused by working and, second, that these are misunderstood if cast as 'accidents'. In this chapter, we have arrived at some sense of the scale of offending related to these deaths and injuries, that is, to discover the extent of safety crimes.

From this exercise, we can draw several conclusions. First, that the scale of such offending is not captured in 'official' crime statistics – if safety offences are crimes then these are separated out from the 'real' crimes recorded by the Home Office. Second, this separation of safety (and, more generally, most corporate) crimes from real crimes is subject to historical dispute, and having important contemporary resonances, remains central to debates around, and obstacles to furthering, safety crimes as a legitimate area of criminological study. Third, we found that in this area (as in many others), studies of the activities of enforcement bodies did not produce reliable estimates of crime *per se*.

That said, we have also insisted that there is enough available evidence – quantitative and qualitative – to justify several further conclusions. Most fundamentally, we have argued that being a victim of workplace violence is more likely than being a victim of 'conventional' violence. But if there is a significant crime problem here, one at least comparable with the crimes of violence upon which the criminal justice system, its political masters, and indeed criminologists focus their overwhelming attention, it remains relatively invisible. Thus, political and popular discussions of crime in general, and crimes of violence in particular, do not include safety crimes. And at the same time, as the previous chapter indicated, when we talk about workplace injuries, it is not to invoke crime or violence, but to imply almost the opposite: namely that these injuries are somehow 'accidents', to the extent that the latter terms remain predominant in academic, policy and popular discourse for referring to those phenomena with which this chapter in particular, and this book in general, is concerned.

If we have established thus far that occupational death and injury represent a significant, yet mystified or obscured, crime problem, then a key, remaining aim of the text is to consider how this crime problem might be reconstructed through law and the criminal justice system. And it is to these tasks that the book now turns as a basis for considering the reform in the treatment and perception of such crimes.

Notes

1 For the sake of completeness, also listed are a series of offences under the heading 'Other', including endangering life at sea, cruelty to or neglect of children, abandoning a child under two years, child abduction, procuring illegal abortion and concealment of birth. See www.crimestatistics.org.uk/output/Page78.asp.

2 Those figures are not made public in official data sources. The figures we use are derived from the monitoring work of CCA (see http://www.corporate accountability.org/manslaughter/cases/convictions.htm).

3 This table is adapted from Table 5, Chapter 2. The absolute numbers are not wholly compatible in the sense that HSE data covers Scotland also. However, the key point of the comparison is with regard to *rates*.

4 In any one year, some of the cases initially recorded as homocide by the police will be reomoved from these records as the police or courts subsequently determine that no homicide has taken place (see Cotton 2004: 4). For the purposes of these tables, the higher – i.e. original – figure has been retained.

5 This figure includes the 172 recorded victims of Dr Harold Shipman.

6 http://www.crimestatistics.org.uk/output/Page63.asp.

7 As Smith and Allen note, 'this is not to suggest that worry about violent crime is unjustified' (Smith and Allen 2004: 1). The point is made here for sake of a crude but, in our view, highly instructive comparison.

8 The original year of publication of Sutherland's classic text *White-Collar Crime*, not to be published in its full, 'uncut' version, until 1983; see Geis and Goff 1983: x–xi.
9 It is in such a context that David Nelken has reminded us that 'the topic of white-collar crime illustrates the possibility of divergence between legal, social and political definitions of criminality – but in so doing it reminds us of the artificiality of all definitions of crime' (Nelken 1994: 366).
10 For a detailed discussion of the debate between Sutherland and Tappan, and the significant implications of their contrasting positions for treating 'corporate crime' as crime, see Pearce and Tombs 1998: 83–124.
11 The source for this data is Health and Safety Commission (2006a) 2005/2006 data are provisional.
12 Based upon data derived from Health and Safety Executive (2005) and Health and Safety Commission (2006a). 2005/2006 data are provisional (p).
13 HSC was established by the HASAW Act 1974, while the HSE, as an amalgamation of existing inspectorates, formally came into being on January 1, 1975.
14 Source of data for 1975–1983 is Dawson *et al.* 1988: 228.
15 Source of data is Health and Safety Commission (2006a) 2005/2006 data are provisional.
16 This is indicative only. For the most part, it is not possible to refer to the percentage of injuries which result in a prosecution, since any prosecution arising from an injury is likely to take place in a subsequent year to that in which the injury itself is recorded, while HSE does not generally make available raw data with which to trace through these linkages. However, see the CCA's analysis of prosecutions following deaths, major injuries, and other categories of incidents (Unison Centre for Corporate Accountability 2002). HSE has since begun to present rates of prosecutions following investigated workplace deaths.
17 We should add a provisos here: while these data cover safety *and health* offences, the data are still utilisable for us since we know that health offences are far less likely to result in prosecution. For example, a CCA analysis of all HSE prosecutions for the three years 1996/97–1998/99 found that there were only 11 prosecutions in total for breaches involving occupational health (cited in Tombs 2004: 172). On this basis, it is likely that the overwhelming majority of the cases that reach court relate to safety rather than health offences. Similarly, the above points do not include reference to dangerous occurences, responding to which appears to be crucial to prevention in HSE's own terms, and those of the Robens system of self-regulation (see Chapter 7). The same CCA analysis referred to above found that in the three-year period under examination, less than 4 per cent (39 recorded dangerous occurences of a total of 927) resulted in a prosecution (ibid.).
18 All figures cited here are from 1998/99. The full analysis can be found in Unison/Centre for Corporate Accountability (2002).
19 Source of data is Health and Safety Executive (2005).
20 There is no simple explanation for this, but some plausible elements of one; see Slapper and Tombs 1999: 45–6.
21 A summary of the cases is included in an appendix (Slapper 2000: 240–68).

Chapter 5

Differentiating safety crime from 'real' crime in law

Introduction

How have safety crimes developed in relation to the mainstream criminal legal process? This is the first of two chapters that focuses upon various dimensions of this question. This chapter discusses the way in which the law has sought to deal with safety crimes, on one hand as 'regulatory' offences, and on the other, as 'crimes of violence'. The question of how the criminal legal process deals with safety crimes immediately raises questions of power. In nineteenth-century Britain, the very first criminal laws that were introduced to protect workers from unsafe working practices, the Factories Acts, proved difficult to implement because of the state's institutional failure to impose criminal status on members of the wealthy class of factory owners. This story provides a key focus of this chapter. The fate of the Factory Acts is also a starting point for this chapter because it illustrates a perennial problem for those who argue for the criminalisation of safety offences: how can a criminal justice system that is designed to deal with lower class offenders be mobilised to punish the relatively powerful individuals and organisations (directors, senior managers and companies) who commit safety crimes?

We approach this chapter from a perspective that views the law not as a naked instrument of power that is always used to control subordinate groups, but as a complex and often contradictory system of rules and practices that ultimately aims to maintain and stabilise the existing social order (Lacey 1994). We therefore take a perspective that is rooted in critical legal studies and neo-Marxist approaches to the law (Grigg-Spall and Ireland 1992) as opposed to instrumentalist approaches (which view law simply as a direct expression of the will of the ruling class) or liberal legal scholarship (which views law as an expression of consensual social values). In order to understand the particular contradictions that lie at the heart of the criminal law, Norrie (2001) argues that we have to explore how the law developed at very particular moments in history and how its shape is derived from the development of social relations at those moments. The development of law as it might apply to corporate offenders in safety crimes is characterised by Norrie as a process of *differentiation* – whereby corporate deviance is strictly segregated – and *assimilation* – whereby corporate deviance is dealt with by adapting the existing criminal law rules that apply to individuals. Differentiation implies the creation and development of a separate sphere of regulatory law and assimilation the re-shaping of corporate liability to integrate corporate offenders into the mainstream criminal process. The criminal law has created a different process for dealing with safety crimes as regulatory offences on one hand and as crimes of violence on the other. This, and the following, chapter is concerned with the legal reasoning behind what we might call a 'bifurcated' model of criminal process. In this chapter, we focus on the ways in which safety crimes are differentiated from 'real' crime, while our next chapter considers some of the details of, and obstacles to, an assimilation of safety crimes into law, a process which grants them the status of 'real' crimes. We begin this chapter by summarising the history of the criminalisation of safety offences, before turning to explore the contradictions and dilemmas that processes of criminalisation create in present day criminal justice systems.

The Factory Acts: the first safety crime laws

The first legislation to intervene in the organisation of factory production came in Britain, in 1802, in the form of the Health and Morals of Apprentices Act, designed specifically to regulate the working conditions of 'Poor Law' apprentices in the textile industry. Then, from 1831 onwards, a series of Factory Acts were passed – regulating the hours and conditions of young workers and women, extending across industries and workplaces of different sizes, until the consolidation of existing legislation in the Factory Act of 1878. Although largely concerned with limiting the working day, the debates that framed the emergence of this legislation were dominated by concerns about the horrific rate of injuries and deaths by 'overwork' of factory workers, and particularly children. Accounts of those conditions published at the time reveal that children were forced to work shifts, sometimes in excess of 24 hours, in overcrowded, filthy and dangerously confined spaces. In Karl Marx's (1887/1954) account of the emergence of the laws aimed ostensibly at curbing the industrial carnage experienced in nineteenth-century Britain, masters of the cotton, silk, and mining industries are revealed as merciless and murderous. Children as young as two were forced into labour in the lace industries and the average labourer in some industrial towns might not be expected to live past their teens.

During this period there existed what Carson has termed a 'logic of enforcement' (Carson 1979: 44–7). If the period between 1802 and 1831 has been characterised by a law that was 'almost entirely ineffectual' (ibid.: 40), with the new law 'contravened on a substantial scale' so as to make it 'a dead letter' (Carson 1981: 136), several somewhat disparate sources of pressure for more effective law and enforcement emerged. Certainly in the debates around the passage and implementation of the Acts in the run up to and then following the 1833 Factory Act (the latter creating the factory inspectorate in the form of four paid inspectors), there was a brief moment during which there was a real impetus to use the criminal law to control the human carnage produced as a result of Britain's industrial revolution – in particular 'to foster a public identity of unambiguous criminality for the delinquent' factory owner (Carson 1974: 129).

A viable class society

Karl Marx's analysis of the emergence of the Factory Acts proposes that the carnage in the factories had created an urgent need for the state to use the law to control factory owners who demonstrated little respect for the rule of law. First, Marx argued that there was an intensification of class conflict between the ruling class and those who worked in the factories. This level of conflict was played out at a level that was threatening the efficient organisation of production. Second, it became clear the industrial system of production could not be trusted to regulate its own rampant physical abuse of labour, and thus threatened the long-term viability of the factory system. The following passage summarises concisely Marx's analysis of regulation as necessary both to dissipate conflict and to check the greed of the factory owners from exhausting its most valuable commodity.

> These Acts curb the passion of capital for a limitless draining of labour power, by forcibly limiting the working day by state regulations, made by a state that is ruled by capitalist and landlord. Apart from the working class movement which daily grew more threatening, the limiting of factory labour was dictated by the same necessity which spread guano over the English fields. The same blind eagerness for plunder that had in one case exhausted the soil, had, in the other, torn up by the roots the living force of the nation. (1887/1954: 229)

The legal protections for workers enshrined in Factory Acts therefore originated in the need to resolve a contradiction inherent in capitalism: the relentless demand for profit in the short term threatened to exhaust the capacity for sustaining profits in the long term. Marx therefore argues that the general impulse to regulate is not simply the result of a consensual or philanthropic decision to make work more humane, but that it is in the very conditions of the factory, the 'dens of misery in which capitalistic exploitation obtains free play for the wildest excesses', that the need for regulation is created (ibid.: 460). Carson later summed this process up by arguing that the Factory Acts helped ensure a 'viable class society' (Carson 1980a).

The introduction of the Factories Acts was central to the process of industrialisation itself. Crucially, through limiting the length of the working day, factory owners were forced to consider ways of extracting greater value from employees in shorter working hours and under ostensibly 'safer' working conditions. The class of factory owners as a whole was impelled to develop technology that would enable the offsetting of costs incurred by regulation. Regulation thus gave capital the impetus to revolutionise production in order to guarantee its own expansion. Paradoxically, then, the Factories Acts hastened the development of the factory system and encouraged the monopolisation of capital (the latter because the increased necessity for a greater outlay of capital forced the smaller factory owners out of business (Marx 1887/1954: 447–8)).

Thus, for Marx, through the imposition of a minimal level of protection for workers, the state played a crucial role in forcing the qualitative leaps in the nature of capital investment and concentration of production that were crucial to the development of the factory system – so effecting the real rather than the formal subsumption of labour. In imposing upon capital conditions of production that were required for what in retrospect was a crucial shift in the development of capitalism, the state helped to establish the conditions of existence and further development of capitalism while apparently acting against the interests of individual employers. It is no casual turn of phrase, then, when Marx refers to this period of early factory regulation (1846–1867) as both 'epoch-making' and constituting the arrival of 'the millennium' (Marx 1954/1887: 268).

Struggles for criminalisation

Throughout his account, Marx takes a great deal of care to underline the contradictory politics at work within the 'bourgeoisie' (ruling class) in the reform process. The compassionate tone of the concerns expressed by wealthy reformers pepper Marx's prelude to the emergence of the Factory Acts. Here we find discontent swelling amongst members of the ruling class, including magistrates who compare children working in the lace trade to slaves in Virginia, alongside establishment newspapers such as

The Times condemning the cruel treatment of factory operatives. The government's own reports to the Childrens' Employment Commissioners on the occupational health of workers in the potteries, match making and paper industries are cited throughout. Those reports document in gruesome detail the physically and morally degenerating conditions of work: vulnerability to disease; low life expectancy; and the mutilation of children, many suffering stunted growth and premature aging.

A succession of Inspectors' reports are quoted for their sustained condemnation of the Factory masters and this condemnation appears to intensify over the years following the Inspectorate's establishment. In particular, *Capital* draws out the different positions adopted within and between different branches of the state. When the Home Secretary eventually appealed to the Inspectorate to treat factory owners who were in breach of the 1844 Act leniently, via a Home Office circular, the Inspectorate refused. The internal divisions within the Factory Inspectorate are also prominent in Marx's account. The least punitive of the four factory inspectors, J. Stuart is singled out for allowing the relay system to flourish in Scotland, whilst in England, the Inspectorate continued 'their legal proceedings against the pro-slavery rebellion' (Marx 1887/1954: 273).

The analysis of the Factory Acts developed by Marx, and latterly Carson, both emphasised the importance of struggles around the *enforcement* of those laws. It is notable that in both of those author's accounts, it was often the case that bitter struggles between employer and factory hand intensified *after* the passage of legislation. For example, the factory owners mobilised against the 1844 Act coming into full force in 1848 with an unremitting lobby of Parliament and appeals to the workforce that their jobs were at risk. Their assault was also directed at the Factories Inspectorate who were denounced as 'a kind of revolutionary commissioners . . . ruthlessly sacrificing the unhappy factory workers to their humanitarian crochet' (ibid.: 270). Marx discussed at length conflicts over the implementation of a 'relay system' by the Factory masters. The relay system, introduced to exploit loopholes in the Factory Acts, enabled masters to circumnavigate the legal limits on the working day by employing workers concurrently across different shifts. The introduction of the relay system represented the manufacturers' 'revolt' against government (ibid.: 397). In response to this revolt the factory workers held protest meetings across Yorkshire and Lancashire, creating a counter revolt so ferocious that the Factories Inspectorate 'urgently warned the government that the antagonism of classes had arrived at an incredible tension' (ibid.: 276). The relay system was finally abolished by the Factory Act of 1850.

Marx is rarely given credit for producing a subtle analysis of state power, but as this discussion shows, his description of the state during this period rejects a crude instrumental notion of the state's relationship with the factory owning class or indeed a crude characterisation of class struggle.

The emergence of strict liability offences

Any zeal for criminialisation in those debates at the birth of the Factories Acts was to fade fast. The courts from an early point had used their prerogative to avoid conviction where it could be shown that the offence was not wilfully or grossly negligent. They also tended to impose the minimum penalties. Just over two thirds of convictions between 1836 and 1842 resulted in a minimum £1 fine (Carson 1979: 50). The discretion being exercised in the courts indicated a major problem of political will to enforce the law.

Proposals for imprisonment as punishment for the most serious offences were rejected in the framing of the 1833 Act. The newly formed inspectorate actually lobbied on occasion for a repeal of the most onerous parts of the law having being advised to co-operate closely with employers so as to make the new law acceptable to them (ibid.: 11–13). The problem the Inspectorate faced was that factory inspectors were being formally

> required to criminalise what was normal within the factory system … to criminalize a body of men not on the periphery of moral life, such as displaced or poverty-stricken workers, but men who were at the centre of the emerging political and social order (Norrie 2001: 85).

Given the routine nature of breaches of the Acts, a response that prosecuted each and every offence would have resulted in a 'collective criminalization which extended far beyond some opprobrious minority' (Carson 1979: 48). With the courts in the main unwilling to embark upon a collective criminalisation of the 'respectable' class of factory owners, the Factories Inspectorate very quickly learned to apply the law selectively. They did this by adopting a system of bargaining and low-level sanctions. Inspectors were given powers to exercise their own discretion, and the Inspectorate powers to devise regulations, which normally imposed an administrative responsibility upon employers (such as keeping time-books, certificates of employees age and so on). Criminal prosecution only came after initial attempts to persuade had failed. Inspectors actually drafted a bill (that was in the event never implemented) to introduce a system of on-the-spot fines to pre-empt criminal proceedings to ease the tricky task of dealing with individuals of high social standing. Despite Hawkins's (2002) failure to consider this history, this is in fact the very origin of 'a strategy of last resort prosecution' (Norrie 2001: 85) with which his work is centrally concerned (see Chapter 7). It is significant that the decision to prosecute came after inspectors had decided that there was some clear intention on the part of the factory owner to commit an offence. This Carson describes as the incorporation of the criminal law concept of *mens rea* into inspectors' decisions to prosecute (Carson 1979: 52–3).[1]

At the same time as the notion of intentionality (and therefore moral blameworthiness) was informally introduced into the investigation process, the legislative reforms of the 1844 Factory Act diluted the normal *mens rea* requirement in criminal process. Specifically, the 1844 Act deemed an employer 'guilty in the first instance' and required the employer to 'prove his due diligence' in defence of any case (Carson 1980b: 164), thus reversing the traditional requirement in criminal law to identify *mens rea* before guilt can be established. This change enabled a second class type of offence to be effectively created within which breaches of the Factory Act would inevitably fall. And *this* offence – since it did not require a state of mind to be demonstrated and carried low-level administrative penalties in the first instance – was more easily constructed as an 'administrative breach' or a 'technical offence', rather than an unambiguously criminal offence. This is essentially why strict liability offences came to be regarded as something other than 'real crimes'. The 1844 Act therefore took a major step towards the concept of 'strict liability' offences which removed the label of criminal from Factory Act prosecutions and thus enabled such prosecutions to be more acceptable to the magistrates who were asked to impose sanctions against members of their own social class. The courts – by ensuring that a separate, lower level of offence was created – assisted in decriminalising breaches of health and safety law in the eyes of the law and thereby legitimated the conventionalisation of those crimes in the workplace.

'Community settlement' and regulation

The analysis developed by Marx, Carson, and Norrie identifies class conflict as the motor of those historical developments. This can be contrasted with liberal perspectives on the development of law which propose that this period was less one of struggle and more simply one of a humanitarian response to exploitation. Changes in the law, from a liberal perspective, arose not from the contradictions created by the system of production and the struggles that arose from those contradictions, but from an almost natural process of finding an optimal level of labour efficiency. For example, in Clarke's overview of the emergence of 'the regulatory society', there is no sense of what this conflict or struggle meant in relation to the regulatory outcome:

> technology, the means to success of industrialisation, brought with it new dangers. In deep mines and factories contracted workers were subject to appalling degradation and exploitation and suffered increasingly from the hazards of new industrial processes. After some opposition, the first Factory Acts were introduced. Their requirements for safer working conditions were respected, on the whole, by the larger manufacturers, but ignored by many of the smaller ones, who could not afford to implement their provisions. (Clarke 2000: 13)

Such a characterisation contains many of the assumptions that infect current debates regarding regulation: that it emerged as part of some unfolding, humanising response to dangers created by 'technology'; that the legitimacy of the law is generally respected by the regulated; and that where regulations are flouted, this tends to be on the part of marginal, less responsible, usually 'small', companies.

Bartrip and Fenn have similarly argued that the Factory Acts emerged from a general consensus that efficient legislation, and its enforcement, rely upon achieving a balance between the benefits of doing business and its 'harmful concomitants' (Bartrip and Fenn 1980a: 178). Safety crimes – because they result from the general momentum of progress – 'differ from other crimes' (ibid.), and thus require a different regulatory response; occasional 'failures' to enforce the criminal law were in fact a function of the constraints within which inspectors were working, so that 'there is no need to look beyond' such constraints 'to explain the evolution of their policy' of enforcement (ibid.: 183); the constraints were 'a reflection of the resources that the community was prepared to devote to enforcement taking into account not only efficiency considerations, but reluctance to erode laissez-faire principles and advance the bureaucratic state' (ibid.: 182); and, as the Inspectorate's practices evolved, it gained 'a more realistic appreciation of what could be carried out in an imperfect world'. (ibid.: 184)

Carson's rejoinder to Bartrip and Fenn identified in the Factory Acts not a 'balance' based upon a community settlement, but an ongoing conflict between pro-regulatory forces and the economic exigencies of the emerging order:

> who exactly is meant to have been making all these rational community choices between lost production levels of 'acceptable' criminality, gains to the community and the like? … Of course the early factory inspectors were operating under constraints; of course they were compelled to balance factors like rigorous enforcement against others such as lost production; but the constraints, I suggest, emanated from something rather more tangible than some ephemeral 'community' sitting down to do its cost-benefit analysis on factory conditions. Rather … they stemmed from a collision between the impetus towards regulation, on the one hand, and basic structural and ideological features of the new order on the other. (Carson 1980a: 190)

The outcome of the implementation and enforcement of the Factory Acts (whether read in the differing historical accounts as 'community settlement' or 'conflict') can hardly be in dispute.

This period of the implementation of the Factory Acts represented a seminal moment in the development of safety law, one in which there were moments of a real possibility of an assimilation of safety crimes into a wider system of criminal law. Through a combination of novel law, emergent forms of enforcement, and the development of a rationale in the courts for treating these crimes in particular ways, that moment was lost, and safety crimes were segregated into a differentiated category of ('strict liability') offences. And this creation of a distinct category of 'regulatory' law in nineteenth-century Britain was to set an enduring pattern which has been followed in many jurisdictions since.

Differentiating liability for safety crimes

The category of strict liability offences that emerged at the end of the nineteenth century facilitated the differentiation of regulatory offences and mainstream crimes of violence. In the UK, this distinction endures, and is made between those offences that are prosecuted as breaches of health and safety law, under health and safety legislation (regulatory offences) and offences that are prosecuted as common law crimes of violence. The most significant category of the latter group of offences are 'corporate manslaughter' offences which are prosecuted in England and Wales using the rules of manslaughter, rules that have been developed to apply to cases involving deaths that result from the involuntary or negligent actions of individuals.

This is not to imply that each of those types of offences can be regarded wholly separately in law. Both are criminal offences. Regulatory offences are defined as criminal offences because, like other crimes, they involve prosecution by the state, they require a pre-determined standard of proof to be demonstrated to the court, and the state imposes a penalty that is linked to the potential social consequences of the offence. Those are the most important characteristics that determine whether a given legal procedure is 'criminal' or not.[2] Offences against regulatory law are governed by criminal rather than civil law. It is this feature of those offences that means that in strict law-book terms we can describe safety crimes unequivocally as crimes. This is important since: 'discussions of corporate crime often ignore altogether the possibility that corporate activity might breach not just the regulatory codes but the criminal law "code" itself' (Wells 2001: 11).

Now, while it is important to recognise that regulatory offences are a sub-category of criminal offences, it is also important to recognise the features that distinguish them from other (mainstream) criminal offences. It is to this task that the chapter now turns.

Inchoate offences

Normally regulatory offences are defined in relation to a breach of law rather than the specific outcome of that breach. Crimes of violence, on the other hand, are defined by the result or outcome of any criminal act or omission, that is, strictly in relation to the harm that has been caused. Thus a corporate manslaughter or homicide charge is only relevant in cases where someone has died, while a company or an individual can generally be prosecuted for a breach of regulatory law regardless of whether some identifiable harm has resulted. Thus, regulatory law 'fails to distinguish those companies which have caused death and injury from those companies which may be unsafe but which have not yet caused harm' (Bergman 2000: 39). Liability that is attributed without any reference to the harm caused is known in legal terminology as inchoate liability.

The removal of the presence of a harm in inchoate offences has been justified by the idea that the primary purpose of the law when it comes to safety crimes is to prevent the harm occurring in the first place, rather than to punish wrongdoing. (Wells 2001: 5–6). This process of disconnecting harm from offence is crucial in reframing the 'criminal' nature of the offence. Some notion of 'harm' has always been a central notion in modern criminal justice systems.[3] If we examine all of the established criminal law textbooks, we find that the notion of preventing harm to individuals and to the public features in any discussion of the founding principles of the criminal law (see, for example, Smith and Hogan 2002: 16–17). The harm principle, as set out by John Stuart Mill ('the only purpose for which power can be rightfully exercised over any member of a civilized community, against his will, is to prevent harm to others'; 1859/1962: 135) was, for early liberal theorists, a precondition for justifying the use of the state's coercive powers over the individual. Harm as a basis for criminal justice intervention remains significant in so far as most criminal acts are in some way connected to the harms that they produce. In cases where the 'harm' is separated from the 'crime', we begin to lose sight both of the gravity that is attached to the offence and, at the same time, of the legitimate basis for state intervention.

Social condemnation of offences under the HASAW Act 1974 may be weakened simply because the offence is not determined by the seriousness of the injury or the harm that results. Because there is no charge of corporate injury or corporate killing under the HASAW Act, it means that the Act generally is not regarded as one designed for punishing serious harms. Rather, it is seen as a statute which aims to prevent a range of unspecified harms before they occur. But those unspecified harms, because they have no label attached to them which draws attention to their status as a 'crime' or to the serious consequences of the crime, are therefore more difficult to comprehend as crimes. Safety offences that use the inchoate mode are described not as 'grevious bodily harm', or 'assault', but instead as 'breaches' of the relevant statute. The effect of all of this is a deepening of the differentiated character of regulatory offences and a reinforcement of the general assumption that they are 'not real crimes'.

Having made this point, it is important to note that the inchoate form does not shape the entire response of the criminal justice system to safety violations. Thus, in the UK, the HSE sets a minimum threshold in the criteria that determine which injuries will be investigated. Some serious injuries are investigated and some are not. According to this policy, all deaths at work are investigated – though this in fact means deaths to employees (see Chapter 2) – as are all amputations, serious multiple fractures, crush injuries leading to internal organ damage, head injuries involving loss of consciousness, burns of more than 10 per cent, blinding, scalping and asphyxiation. The assessment of harm also undoubtedly features in the sentencing process. We will note in Chapter 8, how, through case law, it has been established that the courts should consider higher penalties if the outcome has been a serious injury, ill heath, or a death. Typically the highest fines imposed under the Health and Safety at Work Act are those that involve fatalities. In effect, both regulator (in decisions to prosecute) and the courts here are working within a framework of inchoate offences but are applying harm-based or result-based decisions when it comes to investigation and sentencing respectively.

Corporate liability

Regulatory crimes and crimes of violence are also distinguished by the way in which the criminal law has developed specific procedures for dealing with safety crimes. When safety crimes are processed as crimes of violence (for example corporate manslaughter and corporate homicide cases), the rules of liability closely approximate the normal rules of attributing liability for inter-personal or 'mainstream' crimes of violence. On the other hand, regulatory offences generally imply a method of attributing liability that has been designed for the purpose.

We have already noted the criminal justice system's pre-occupation with *individual* culpability for safety crimes. The central problem with the criminal law's methodological individualism for our discussion here is that responsibility for safety crimes can rarely be meaningfully attributed to one or several specific individuals, as is the case with inter-personal crimes. Safety crimes often result from a series of decisions made at different levels of the organisation. In the case of Piper Alpha (see Chapter 1), the decision to reduce the maintenance programme on the platform and at the same time push the production system beyond its capacity was crucial to the genesis of this safety crime. The decisions that led to this state of affairs were made at various levels in Occidental's corporate hierarchy. They were also directly the result of poor management–workforce relations and the lack of any formal worker representation which could force management to respond to workers' complaints about safety on the plant (Woolfson *et al.* 1996; Whyte 1999a; Tombs 1990). Thus the killing of 167 workers on the Piper Alpha platform cannot be understood in terms of the result of a decision of any one individual. The cause of the Piper Alpha disaster, as is often the case with other safety crimes, can only meaningfully be understood in terms of its organisational causation or production.

There are a variety of types of employing organisation that can be implicated in safety crimes: corporations, government and public sector organisations, charities and associations and various other forms of unincorporated organisations. The constitution of public sector organisations is generally governed by public law and the constitution of private sector organisations by corporate law. However, most organisations have similar safety responsibilities in regulatory law. In general, regulatory law applies to both public and private sector organisations.

For Glasbeek (2005), however, there is an important distinction to be made between corporations and other types of organisation. What makes corporations different is that they are constituted, socially and legally, for one primary purpose: to maximise profits from commercial activities. This means that law reformers must recognise corporations as a special case. Corporations will make profits in whichever way they can and this includes cutting corners and exposing workers to risk (unless there are restrictions in place which can limit such exposures). Law reformers, however, tend to work within a consensus perspective (see Chapter 7) and work on the assumption that good safety is in the best interests of all organisations. For Glasbeek, lumping all organisations together in one category supports a very specific approach to criminalisation. The primary aim is to find reforms that will deal with the complexity of organisations rather than those that seek to change the institutional basis of the corporation and limit the murderous effects of the pursuit of profit. This is a point that has resonance with Norrie's (2001) argument on the removal of *motive* from the criminal process. Motive – in this case, profit motive – is irrelevant to the process of establishing liability for a crime. By obscuring the profit motive, the law 'serves another unmentioned purpose: it supports the status quo in respect of occupational health and safety regulation' by preserving the consensus view that safety crimes are generally 'accidents' rather than the avoidable consequences of an industrial profit-maximising process. Thus, the consensus view of safety crime and how to regulate it (see Chapter 7) is preserved: 'when it comes to the risks of any for-profit activity, the employer is in the best position to know what they are likely to be and how they could be minimised' (Glasbeek, 2005: 25).

Having made the point that there is good reason for corporations to be treated as a special case, it remains the case that safety crimes – albeit in different forms and for different motivating factors – are committed by a range of different types of organisations. In the UK, part of the debate on the form that a new offence of corporate manslaughter might take has revolved around the extent to which crown bodies (institutions which represent the state such as government ministries) should be held responsible for criminal offences. It is the desire to ensure that crown bodies and non-incorporated bodies do not remain exempt from prosecution which has led to calls for the law to be reformed to allow all types of organisation to be held accountable. The implication of Glasbeek's argument, however, is that the law must be sensitive to the constitution of different organisational forms and therefore be sensitive to the underlying reasons behind the organisational production of safety crimes.

The argument that profit-making corporations should be treated as special cases is supported by the fact that, as we shall see in the following chapter, corporations are able to fall back upon a legal principle which protects them from liability known as 'corporate personality'. Indeed, it might be said that the problems that have been historically created by the process of assimilating corporations into the mainstream criminal law stem from the special status that corporations are given; a status which enables the corporation to own property formally and to make profits for the individuals that in reality own and control them, *without* those individuals incurring full liability for any harms that they may cause vicariously through the corporation. In other words, corporations present specific problems for the criminalisation of safety crime.

Principles of corporate liability

A key period in the development of strict liability offences – outlined above – was undoubtedly the struggle to implement the Factory Acts. However, this is not to imply that the concept of strict liability was only developed for the purpose of dealing with safety crimes. As Wells (2001: 68) noted, the concept in the ninteenth-century was 'more commonly found in the area of morals, minors and drink'. Yet the concept does have a property that makes it particularly amendable to corporate offending. In strict liability offences, it does not matter if the accused is an individual or an organisation: the absence of *mens rea* ensures that either the organisation or the individual (or both) could be held liable where relevant causation is shown. Normally, however, it is the corporation that is prosecuted for safety crimes. In this sense, the legal category of strict liability has allowed law enforcers to avoid a central dilemma in debates around the criminalisation of safety crimes: should the law focus on the organisation or on an individual within the organisation? Historically, the development of differentiated forms of liability for safety crimes has focused predominantly upon how to attribute liability to the organisation rather than the individuals who contribute to organisational decision making; we shall see in the following chapter the extent to which regulatory offences in the UK that apply to directors and senior managers are ill-designed for this purpose.

Methods for imputing liability to corporations generally use mechanisms that impose criminal responsibility for the conduct of employees of the corporation upon the corporation itself. Although the meaning of the term changes according to the context in which it is applied, 'vicarious liability' is normally used to imply that the corporation's liability can be imputed using the conduct of one or more of its employers. A pure form of vicarious liability would enable the actions of any employee at any level of the organisation to be imputed to the corporation. Some selective forms of vicarious liability restrict this mechanism only to a specified group within the organisation. It is often said that US law uses a system of vicarious liability. Normally what is being referred to here is a very particular form of vicarious liability known as *respondeat superior* which applies in the US Federal Courts. Under the doctrine of *respondeat superior*, corporate liability can be triggered by the criminal conduct of any employee of the corporation, as long as the conduct occurred within the scope of their employment (that is, the employee was conducting a work-related activity) and the conduct was intended to benefit the corporation (see Gobert and Punch 2003: 55–9; Wells 2001: 132–6). There are two main problems with this doctrine. The first is that the relative ease with which *respondeat superior* can be used to attribute liability means that the courts have developed a relatively low level of punishment, in other words, a similar effect to the differentiation of strict liability cases. Vicarious liability cases in the US courts are regarded as part of a system of rules that do not fall squarely within the parameters of the criminal law. The second is a more conceptual problem that raises questions about whether vicarious liability can fully capture the *corporate* nature of the offences it is used to prosecute. Although *respondeat superior* represents a relatively broad method of attributing liability, it still relies upon identifying the conduct of an individual who can be said to have triggered corporate liability. This may, particularly in complex organisational structures, be difficult, in spite of the fact that this individual can be at any level of the organisation. Vicarious liability-based systems therefore do not take us closer to a meaningful attribution of the offence to corporations because they constitute a form of liability in which the individual remains central to the establishment of fault.

This latter problem reminds us that when thinking about corporate crime it is always necessary to understand the relationship between the offence, the corporation, and the individuals who make up the corporation. Understanding these relationships accurately is necessary if we are to reverse a situation in which corporations continue to escape criminal responsibility (Kaye 1992). At the beginning of this chapter, we asked: how can a criminal justice system that is designed to deal with lower class offenders be mobilised to punish the relatively powerful individuals and organisations (directors, senior managers and companies) who commit safety crimes? At the end of the chapter, the question has become rather more complex: how do we use a body of law designed for individual offenders when we are dealing with a form of crime that very often can only meaningfully be attributed to collective action and collective decision making, perhaps at multiple levels in the corporate organisation?

Conclusion

The origins of a differentiated system of law for safety crimes can be traced to the moment at which the criminal courts saw the punishment of the wealthy or the ruling class as somehow beyond their remit. The process of law reform and the process of law implementation remains guided by a logic of social ordering – of mediating over conflict in a way that will not disturb the status quo. We cannot escape the fact that a necessary, key logic of capitalist social orders is to maintain the predominance of capital – and its institutional form, the corporation – in social relations. It is this fact above all else that shapes the criminal law's relationship to safety crimes (Slapper 1993; Glasbeek 1989).

The process of differentiating law identifies an enduring paradox in the criminalisation of safety crimes. In the case of the Factory Acts, a new principle of liability that emerged after protracted struggles around the implementation and enforcement of the criminal law enabled a second class of crimes to be created which would never be able to attract the same level of moral opprobrium or condemnation attached to other crimes of violence. The effect of 'strict liability' in the case of the Factory Acts was not to make the criminalisation process any easier, but to differentiate safety offences as a distinct category of crimes. This process of *differentiation* is deepened by the creation of separate systems of policing and punishment that apply to safety crimes. Because strict liability offences are not regarded as 'real' crimes, the public relations consequences of regulatory offences and the relatively low-level sanctions imposed by the courts (if those cases actually reach the courts) are absorbed rather more easily by convicted corporations and individuals.

There is a very crude principle of law reform in operation here: if you make the route to establishing liability easier, then it is the instinct of the court to interpret a lower degree of seriousness and therefore attach a less severe punishment to the offence. Lower thresholds of liability imply a relative ease with which prosecutions for safety crimes can be secured. However, this possibility of easier criminalisation is offset by the lack of resources that investigating and prosecuting authorities can draw upon, the resort to less punitive pre-court sanctions, and consequent low rates of prosecution for safety crimes. This effect highlights an inherent contradiction in the legal rules that have been developed to hold corporations accountable. Put simply, a process of decriminalisation has occurred at the same time as the rules of prosecution being developed in a way that makes it easier to convict. If we are effectively to criminalise corporate crimes, then the most obvious conclusion to draw from this contradiction is that we should seek to assimilate safety crimes into the system of criminal law. Yet, developing the law in this direction has also been beset by a similarly perplexing contradiction: if we are to fully criminalise safety crimes, does this mean that we have to assimilate safety crimes into a system that has not been designed for the purpose? It is to the key legal issues raised by this question that we now turn.

Notes

1 In criminal cases, the latin phrases *actus reus* and *mens rea* are used to describe the basic elements of a crime. For a crime to have been committed, there needs to be presence of criminal conduct or the physical cause of a criminal event (the *actus reus*) and an identifiable individual with a criminal state of mind (*mens rea*). Actus reus is often referred to as the 'physical' element of the crime and *mens rea* as the 'mental' element of the crime. The literal translations of *mens rea* is 'knowing mind'. The term is used in a legal sense to mean that intent or knowledge on the part of the offender is required before a crime can be said to have occured.

2 By contrast, civil law cases involve no prosecution, but are in essence applications to the court by private citizens for adjudications to be made on a balance of probabilities. There is no 'accused' in a civil court case; court decisions are normally made to establish liability on one side or the other, between the plaintiff (the person applying to the court for a settlement) and the appellant (the person who defends him/herself against the case made by the plaintiff).

3 Even if, far from dealing with the most pressing public harms, the crime that is generally dealt with by the apparatuses and technologies of crime control involves a few, highly individualised harms (Hillyard *et al.* 2004).

Chapter 6

Assimilating safety crimes

Introduction

In the previous chapter, having introduced Norrie's (2001) distinction between the processes of *differentiation* and *assimilation*, we considered the historical basis for the fact that safety crimes are not regarded as real crimes in law. Thus we argued that organisations and their directors are rarely exposed to the same criminal controls or level of punishment as relatively lower status individuals (Sanders 1985). Offenders who are responsible for committing safety crimes are generally regarded as socially useful and respectable, and often – because of their status, wealth and other resources – enjoy some measure of influence over public policy decisions. Moreover, employing organisations, and their individual directors and senior managers, are insulated by a separate set of rules that guide criminal prosecution for safety crimes. In this chapter, we turn to examine a process of assimilation, whereby corporate deviance is dealt with by adapting the existing criminal law rules that apply to individuals. On the face of things, a process of assimilation allows corporate crime – including safety crime – to be treated as 'real' crime. Having said this, the process of assimilating crimes of the powerful more generally is one that is rare in any criminal justice system. Nonetheless, it is a process that deserves particular explanation in this text in order to consider the potential for safety crime to be dealt with effectively by a legal system that is designed to deal with the crimes of the relatively powerless.

The first, and perhaps most fundamental, problem with the assimilation of safety crimes into the mainstream criminal legal process is that the criminal law in the western legal jurisdictions has been designed with lower-class individual offenders in mind. The rules of *mens rea* that have been developed in Western jurisdictions tend to involve several levels of culpability. This is necessary because it is not always the case that individuals can be said to have clearly intended the consequences of their actions. In safety crimes, it is not often the case that an act or omission is intended to injure or kill. Often in those cases, risks to workers or the public might be created without any clear intention to kill or maim. Cases of safety crimes therefore normally imply a lower level of culpability than intent. Safety crimes are more likely to result from the *reckless* or *careless* disregard for the consequences of taking risks.

In most Western legal systems, this distinction can be characterised as the essential difference between 'murder' and 'manslaughter'. The offence of murder normally implies intent to kill, whereas manslaughter normally implies that a death has arisen from reckless, unintentional conduct. In such cases, the test for criminal liability is normally that the risk was foreseen or was foreseeable. In other words, the risk should have been foreseen by any reasonable person. In English law, the test is that risk must have been serious and obvious enough for it to have been appreciated by the accused and that the accused was capable of appreciating the risk.

However, if manslaughter has been a frequent analogy drawn upon by academics and safety campaigners alike – and indeed has been one route that the criminalisation of safety crimes has followed – there are other properties of safety crimes that make them appear rather more like murder than manslaughter. It is possible to argue that, although their consequences tend to be unintended, safety crimes are often committed under conditions that of greater pre-meditation and planning than many murders. As we have seen in Chapter 1 (and we develop this discussion in the conclusion), the typical causes of safety crimes can often be linked directly to the demands of cost-driven managerial regimes. Thus aggresive managements, pressures upon employees through time or resource constraints, inherently dangerous ways of working established as standard operating procedures, inadequate training, poorly designed or maintained equipment, the absence or lack of robustness of basic safety equipment, and the use of casualised, overworked or under-qualified staff can all be understood as consequences of profit-motivated decisions made, or cultures set, by managements. In Chapter 1, we saw how decisions to cut the maintenance budget on Piper Alpha, to decrease the harbour turnaround time in P&O channel sailing or to put untrained staff into high risk work, as in the case of Simon Jones, were all causal factors in those crimes. In those terms, they were all crimes that can be traced back to the carefully planned decisions of accountants and managers. Notwithstanding the stereotypes of crime fiction, most murders do not tend to arise from premeditated, calculating plans, but are often the result of a pattern of violent interactions, commonly between individuals who know each other. Safety crimes on the other hand are, by definition, committed in cold blood and often are committed in order to advance financial goals or maintain profitable systems of production. We therefore need to think through carefully the distinctions between safety crimes and other types of crimes of violence – distinctions that we began to draw out in Chapter 3. At the very least, however, it is perfectly plausible to suggest that the differentiated treatment of safety crimes is not a consequence of properties inherent to such offences which render them less blameworthy. As we noted in Chapter 3, following Reiman, it may be reasonable to attach a higher degree of moral blameworthiness to such crimes. In other words, it is the concepts and categories that are used in mainstream law, and the distinctive approaches that have been adapted to safety crime, rather than the characteristics of the offence *per se*, that stand in the way of criminalisation of safety offences.

Acts and omissions

In cases of safety crimes, the concept of 'knowledge' is likely to be defined as an individual's awareness of the consequences of taking a particular risk in the workplace. So if a manager knowingly allows employees to work from a height without a safety harness, then it might be argued that sufficient *mens rea* is present to allow the incident to be defined as a crime. But a safety crime rarely arises directly as a result of a positive decision. Safety crimes are more likely to result from indifference, and the failure or refusal to provide adequate protection from harm. In the case of workplace safety, offences tend to be caused by failures to comply with the minimum legal requirements, where the deadly effects of those are entirely predictable. Safety crimes may often be the result of management failure to do something they should have done to protect workers and the public. In other words, safety crimes tend to be produced by acts of *omission*, rather than acts of *commission*.

A criminal omission is generally more difficult to pin down than a criminal act. It may often be much harder to identify the lack of action on the part of a particular manager or boardroom, than it is to identify an act that creates a serious risk to workers or the public. Of course there are some cases where omissions are clear. In the Lyme Bay case, it was relatively easy for the prosecution to show that the accused, Peter Kite, was aware of the consequences of his inactions. In December 1994 the company concerned, OLL Ltd, became the first company in English legal history to be convicted of homicide. Peter Kite, its managing director, also became the first director to be given an immediate custodial sentence for a manslaughter conviction arising from the operation of a business. Both Kite and the company were found guilty on four counts of manslaughter arising from the death of four teenagers who drowned in Lyme Bay while on a canoing trip organised by OLL. Kite was sentenced to three years imprisonment, and the company was fined £60,000. But the case was arguably atypical of corporate homicide scenarios. In this case, there was evidence that the risks had been made clear to him by employees and that he had chosen to ignore those risks. The key evidence presented to the court was a letter from a previous employee warning Kite of the serious risks of loss of life created by company procedures. Without this evidence, the court may not have found it easy to convict.

By contrast, in the case of the Transco Larkhall explosion in 1998 (see Chapter 1), the neglect of the gas pipeline system took place over a number of years and the failure to take action to upgrade the system was identified as the responsibility of a number of individuals in a number of committees and working groups. Ultimately, then, the case against the company did not proceed because the Crown did not identify one individual with sufficient authority and responsibility for the offence to trigger corporate *mens rea*.

These points being made, we must emphasise that there is no necessary antagonism between the criminal law as it is currently organised, and the criminalisation of omission. As Jefferson (2001: 132) has noted, if 'English law traditionally does not always hold a person guilty for failing to act', there 'are exceptional cases', and these 'exceptions have grown in recent years'. Thus, discussing a series of 'established exceptions', Jefferson notes the following, of particular interest to, and relevance for, us: failure to perform a duty imposed by contract; failure to perform a duty imposed by law; and the unwitting – that is, where there is no *mens rea* – creation of a dangerous situation, since this creates a duty to put that situation right (ibid.: 132–41). Moreover, he notes that, 'Since death may be caused by omission, so also may the lesser statutory offence of causing grievous bodily harm with intent' (ibid.: 135).

Now, there is no doubt that some of the specific claims made by Jefferson may be contested, the relevant case law being open to differing interpretations. But what his discussion absolutely clarifies is that there is no necessary technical impediment in criminal law to dealing with crimes of omission. In turn this indicates that there is space for judicial development and/or legal reform, so that the lack of such development or failure to reform is less a technical or legal issue, and more a political or ideological one.

Corporate criminal personality

In regulatory law, as we have seen in the discussion of the history of the Factory Acts in the previous chapter, the test of *mens rea* was removed to produce a new principle of liability, thereby differentiating corporate from individual offending. The effect of this legal invention of strict liability was to sidestep the thorny issue of assimilating safety crime into the system of criminal rules that apply to individuals. The convenience of strict liability was that it removed the 'mental element', or the need to retain a concept of intent, and therefore it could eventually be applied relatively unproblematically both to corporations and to individuals. But strict liability was not to remain as the only basis for attributing liability in safety crimes. As case law developed throughout the twentieth century, the courts developed a method of using the rules of individual liability in relation to corporations. In essence, what happened was that the criminal law principle of *mens rea* was reshaped to enable liability to be imputed to organisations and collective entities. As part of this process of reform, the law contrived a way of manufacturing a 'corporate criminal personality'.

The corporate criminal personality is analogous to, but significantly more complex than, the legal personhood ascribed to the corporation in corporate law. In the latter, the concept of limited liability allows for corporations to be granted a separate 'personality' which is distinct from the individuals and organisational processes that comprise the corporation. Corporate personhood allows the financial liability of shareholders and managers to be limited because those liabilities are deemed to be held by separate legal entity in the 'person' of the corporation. Limited liability thus disconnects the corporate person from the persons that make up the corporation. The principle of limited liability is therefore the defining feature of the modern corporation. Limited liability effectively reduces exposure for investors in the event of financial losses being incurred by the corporation. The principle of limited liability which protects corporations in Western systems of law ensures that owners (normally shareholders) are not responsible for the financial liabilities of the corporation (Glasbeek 2002; Spencer 2002). If senior managers take a series of decisions that plunge the company into debt, managers or owners of the company usually will not be held liable for that debt. Investors can only lose the shares they invested in the first place, nothing more. In legal terminology, this protection for owners is known as the 'corporate veil'. In human rights law, corporations can claim the same 'rights' (to life, privacy, freedom of expression, and so on) as can individuals. The principal effect of the fictional legal form of the corporate person is to protect the property rights of capital and to encourage the reproduction and accumulation of capital by corporations.

At the same time as corporate personhood is held in company law to be a fixed legal state in which the limited liability corporation is protected by the same rights as any individual, the criminal law constructs a parallel but very different conceptualisation of corporations. The corporate *criminal* personality is brought to life only after an individual within the organisation is first implicated in the offence. This legal principle is captured in the relatively new doctrine of corporate *mens rea.*

The identification doctrine

Whilst *mens rea* can be said to be a relatively new doctrine, the notion of collective or group liability is an old one. In twelfth-century England, communities were held collectively liable for murder. As Wells notes, in medieval thinking, the individual was regarded as being inseparable from group and community membership. Now, the individual has been abstracted 'from the complex of relationships, such as nationality, class, or occupation, by which she was previously defined' (2001: 72). The enlightenment notion of the autonomous, rational self-interested individual therefore precipitated a shift in the way we think about both individual and collective liability. This methodological individualism – which became embedded in eighteenth- and ninteenth-century English law – has had profound consequences for how the courts developed a concept of corporate liability.

The concept of corporate *mens rea*, which evolved in the English courts and was followed by other common law jurisdictions throughout the twentieth century, assumes that a corporation can not be held liable for a crime unless an individual within the organisation with sufficient knowledge of the offence and with the necessary responsibility and authority in the organisation can first be identified. In other words, corporate *mens rea* depends upon identifying a particular 'controlling mind' within the organization that can be said to be responsible for the offence. There is a long and complex case history of the evolution of corporate *mens rea* and we are therefore wary of over-simplifying the issues (see Wells 2001; Slapper and Tombs 1999: 26–35; and Gobert and Punch 2003: 59–69). Having noted this as a word of caution, a good illustration of the origins of principle applied for determining who are the people representing the 'controlling minds' of the corporation is summed up in a dictum of Lord Denning's in the 1957 case of *H.L. Bolton (Engineering) Co. Ltd.* v. *T.J. Graham & Sons*:

> A company may in many ways be likened to a human body. It has a brain and a nerve centre which controls what it does. It also has hands which hold the tools and act in accordance with directions from the

> centre. Some of the people in the company are mere servants and agents who are nothing more than hands to do the work and cannot be said to represent the mind or will. Others are directors and managers who represent the directing mind and will of the company and control what it does. The state of mind of these managers is the state of mind of the company and it is treated by the law as such.[1]

Latterly in a key case, Viscount Dilhorne in the case of *Tesco Supermarkets Ltd.* v. *Nattrass* defined the controlling mind:

> … who is in actual control of the operations of a company or of part of them and who is not responsible to another person in the company for the manner in which he discharges his duties in the sense of being under his orders.[2]

The 'identification doctrine' can be understood as an extension of the notion of 'corporate personality' since it establishes the corporation as a separate entity from the directors and managers that make decisions on behalf of the corporation. Yet it departs from 'corporate personality' in so far as it establishes a complex and often difficult-to-establish route to establishing the organisation's criminal state which, unlike the concept of corporate personality, rests on the identification of an individual's acts or omissions. The key problem that has been experienced in a succession of cases is the difficulty of actually identifying one individual who can be said to constitute the controlling mind of the organisation. This individual must also be shown to have been grossly negligent and to have been aware that the risk of causing death is both serious and obvious. Although in a series of cases in the 1980s the courts failed to make the basis for prosecution clear, it seems that this test remains a tough one (for a discussion of the complexities of this debate, see Hartley 2001; and Bergman 2000).

Referring back to the Lyme Bay case outlined above, the relatively small size of the company OLL Ltd made it reasonably easy to identify an individual who can be said to have been the 'controlling mind'. The case for the prosecution was bolstered by the evidence of a letter from the former employees that indisputably made the managing director aware of the risks in question; serious risks that, as the prosecution was able to show, were not subsequently addressed with any urgency. Between the OLL Ltd prosecution and April 2006, six more companies have been convicted for corporate manslaughter. Most of those convictions have been against very small companies.[3] As Sullivan (2001: 33) has noted, the law as it stands 'ensures that, *de facto*, liability is confined to small companies' (see also Dunford and Ridley 1996). Here we find an increasingly glaring irony: while the very size and complexity of large organisations have been key features in *producing* disaster, those same features have rendered the identification of a controlling mind, and thus corporate prosecution, virtually impossible (Tombs and Whyte 2003a).

The most notable challenges to the identification doctrine came in the P&O case (see Chapter 1) and then in the prosecution of Great Western Trains over the Southall train crash, where, in 1997, six people died and some 150 were injured after an InterCity passenger train ploughed into a freight train. In the latter case, the Crown tried to argue that the doctrine of identification no longer applied and that it was possible to consider the conduct of the 'company' as a whole rather than the conduct of an individual 'controlling mind'. The trial judge however ruled that:

> There is, I accept, some attraction in the fact that gross negligence manslaughter, involving as it does an objective test rather than *mens rea* in the strict sense of the expression, is in some ways closer to statutory offences of the kind in the cases relied on by Mr Lissack than it is to the ordinary run of criminal offences. But I do not think this is a good reason for no longer having to look for a directing mind in the company to identify where the fault occurred. In my judgment it is still necessary to look for such a directing mind and identify whose gross negligence it is that fixes the company with criminal responsibility.
>
> Accordingly, I conclude that the doctrine of identification which is both clear, certain and established is the relevant doctrine by which a corporation may be fixed with liability for manslaughter by gross negligence.

As a result of this ruling, the Attorney General asked the Court of Appeal for a clarification of the law on this point. In Attorney General's Reference No 2/1999, the court held that the Trial judge was correct and that 'the identification principle remains the only basis in common law for corporate liability for gross negligence manslaughter'. In other words, at least as we write,[4] *mens rea* remains a pre-requisite for corporate manslaughter cases to proceed.

The corporate *mens rea* approach was developed in order to allow organisations to fit within a body of law that, as we have seen earlier in this chapter, is underpinned by methodological individualism. This attempt to resolve a contradiction thrown up by the criminal law has also created a new internal contradiction: that organisational prosecutions are ultimately dependant upon the case against an individual. As Tim Kaye has remarked, ' … it is a very strange form of legal personality which says that a person is legally responsible only when someone else is guilty as well' (1992: 354; see also Neocleous 2003: 84).

The principle of aggregation

Safety crimes, like many corporate crimes in general, are rarely imputable to the actions of one individual or a group of individuals acting alone. The principle of aggregation presumes that corporate *mens rea* can be imputed to an organisation by combining the acts or omissions of several different agents of the corporation across managerial levels of the organisation and across time. Claims that a principle of aggregation might apply in law have been raised in arguments heard at two landmark trials, the P&O case in England and the Transco case in Scotland.

Consistent with the doctrine of identification, in order to establish that P&O itself was guilty of manslaughter, the prosecution had to first prove that one of its controlling directors or senior managers was guilty of manslaughter. When the prosecution against five senior employees collapsed, the case against the company inevitably went too. Further restrictions on the criminal responsibility of corporations were set out in a High Court ruling which in effect precluded the aggregation of fault from several directors as being sufficient to incriminate the company. On this matter, Bingham L.J. ruled that:

> Whether the defendant is a corporation or a personal defendant, the ingredients of manslaughter must be established by proving the necessary *mens rea* and *actus reus* against it or him by evidence properly to be relied on against it or him. A case against a personal defendant cannot be fortified by evidence against another defendant. The case against a corporation can only be made by evidence properly addressed to showing guilt on the part of the corporation as such.[5]

If this sounds complex, that's because it is. But we can cut through the circularity of the argument to some extent if we retain some notion of a simultaneous process of *assimilation* and *differentiation*. What Lord Justice Bingham is arguing here is that if we apply the normal rules of criminal process, the courts would not usually allow the actions of one individual to contribute to another's guilt. By doing so, the precedent that has followed this case has effectively fixed the notion of corporate personhood firmly in the arena of criminal law. In other words, he is therefore assimilating very narrowly the corporation into the criminal process. Because often corporate killing involves aggregation, for reasons indicated above, the precedent set here has ensured that a great many cases will not be heard in front of a criminal court. This ruling shows therefore that attempts to assimilate corporations into a system of law designed for individuals can actually have the effect of differentiating safety crimes. This is the paradox that the law has created with the doctrine of identification, one which others have pointed out, stems from Bingham having 'knotted himself' in a circular argument (Slapper, personal interview cited in Hartley 2001). It is this paradox that, as Kaye (1992: 349) observes, makes the criminal law look decidedly 'schizophrenic' when it is applied to the corporation.

The key advantage in the principle of aggregation is that individual states of mind can be added together in order to produce a corporate guilty mind. The P&O ruling following the sinking of the *Herald of Free Enterprise* denied the legal integrity of this principle and a very similar endpoint was reached by the court in the Transco case (see Chapter 1). In the latter case, the failure to review and replace dangerous piping by the firm took place in several meetings at different levels of the organisation which spanned a number of years. The prosecution's novel argument that such decisions and non-decisions could be aggregated across time to impute corporate liability was also denied by the court. The case was widely known as 'Scotland's Herald of Free Enterprise', a title which has an unintended, but pointed, double meaning.

The P&O and Transco cases highlight important contradictions in the law as it applies to corporations. First, they highlight a general paradox. Organisational crimes are very often more easily committed because of the diffusion of responsibility and blame that occurs in large organisations. Complicity in crimes of a bureaucracy or an organisation is much more amenable to techniques of neutralisation on the part of individuals involved, quite simply because they may be able to point to multiple causes at several levels of the organisation. The very same features of an organisation that make safety crimes more likely, then, are the features that prevent the perpetrator being held culpable for manslaughter/homicide in a criminal court. Second, these cases also denied the possibility that the actions of different individuals at different stages of the decision-making process could be aggregated to produce organisational fault. Yet, as we have seen already in this chapter, there is a contradiction in company law, whereby corporations are constructed in a very specific way that enables the various components of the organisation and its decision making processes to be considered as a singular whole. The P&O and Transco rulings reveal an inherent bias in the law's concept of corporate personhood. In those cases, the idea that the corporation could be regarded as a integrated decision-maker for the purposes of criminal conduct was effectively rejected. The courts dispensed with the principle of aggregation at the point when its application would implicate the company in a serious crime. The law, in other words, grants the corporation a singular, human-like identity that protects it from the legal liabilities that human beings cannot escape – yet this singular identity apparently does not apply when the corporation is being held to account for its crimes.

Generalising corporate liability

To this point, we have been examining how the criminal law in England and Wales[6] has developed a form of corporate liability for manslaughter. Yet we might extend our deliberations beyond deaths, to consider how the law might proceed against corporations for crimes of violence resulting in injuries short of death.

Now there is no reason, theoretically at least, why corporate *mens rea* could not be applied to a whole range of crimes of violence caused by corporations (such as corporate endangerment or corporate assault offences; Bergman 2000). In practice, of course, prosecuting authorities have not been minded to bring such cases. If there is no doctrinal legal reason for this, then the explanation of such a limited approach to assimilation must lie somewhere beyond the legal process. In England and Wales, there has been no sustained public debate about corporate violence beyond the high profile 'disaster' cases.

The idea that corporations can be held liable for all criminal offences is by no means a new one. In the 1699 edition of George MacKenzie's *Law and Customes of Scotland in Matters Criminal*, a text that is recognised as the first attempt to codify the criminal law in Scotland, the author, who was Scotland's most senior jurist of the time, discusses whether a collective body of people or an 'incorporation' can commit a crime or not. His conclusion was that, under certain circumstances, incorporations could indeed commit crime and could be punished for those crimes, and that 'rulers' of the incorporation – the individuals responsible for the incorporation's crime – should be likewise punished. Some have speculated that had it not been for the 1707 Act of Union, Scots law under MacKenzie's influence would have been codified. Had this transpired, it is possible that a general principle of corporate liability might have featured in Scotland's criminal code since the early eighteenth century. Of course this did not transpire. A more recent Draft Criminal Code published by the Scottish Law Commission (Clive *et al.* 2003) notes that 'the existing law on corporate liability is not satisfactory' and argues that corporations as 'legal persons' should be liable for a range of offences in the criminal code.

Some jurisdictions, such as France, have sought to introduce corporate criminal liability based upon rigid principles of corporate *mens rea* (Wells 2001). In Canada, the C-45 Bill, introduced by a reform process triggered by the Westray disaster in which 26 miners died, created a new way of attributing criminal liability for organisations. The reformed Criminal Code in Canada allowed 'organizations' to be prosecuted for any of the criminal offences set out in the Criminal Code – not just involving deaths.

It covers, for example, the offence of 'causing bodily harm by criminal negligence' as well as financial and other offences which go beyond the realm of safety crimes. At the time of writing, the reforms in Canada are relatively new and have yet to be fully tested by the courts.

A particularly interesting attempt to create a novel form of corporate liability known as the 'corporate culture model' has been developed recently in Australia. Under the Australian Federal Criminal Code Act (1995), liability can be attributed to a corporation if its board of directors or a senior manager intentionally, knowingly or recklessly carried out the relevant conduct. Intention, knowledge or recklessness can be also attributed to a corporation where it is established either that a 'corporate culture existed within the body corporate that directed, encouraged, tolerated or led to non-compliance with the relevant provision' or where the corporation failed to create and maintain a corporate culture that ensured legal compliance. Corporate culture is defined in the Act as 'an attitude, policy, rule, course of conduct or practice existing within the body corporate generally or in the part of the body corporate in which the relevant activities takes place'.[7] In the corporate culture model, liability can be imputed where the individual who committed the unlawful act reasonably believed that an authoritative member of the corporation would have authorised or permitted the commission of the offence. An authorisation or permission can be established by proving that, on the one hand, the board of directors, or a high managerial agent of the body corporate, directly authorised, or implied authorisation, or permitted the commission of the offence. On the other hand, the authorisation or permission can also be established by proving that a corporate culture existed within the body corporate or that the body corporate failed to create and maintain a corporate culture that required compliance with the relevant provision. Effectively, then, in the corporate culture model, there is no need to show that any one person acting within the corporation was negligent. Negligence can be found in the conduct of the corporation viewed as a whole. This is reflected in Section 2 of Section 12.4 of the Act which states that 'fault element may exist on the part of the body corporate if the body corporate's conduct is negligent when viewed as a whole (that is, by aggregating the conduct of any number of its employees, agents or officers)'.[8]

It is also worth drawing attention to recent, novel attempts in Italy to develop a model of corporate criminal liability – albeit one generated in the context of efforts to pursue financial rather than safety offences. In March 2002, legislation was passed to create two types of corporate liability. One invokes crimes committed by the head of a corporation, and thus is not far removed from the principle of identification. Much more significant, however, is a second, 'innovative element', which has established the concept of 'structural negligence' (Gobert and Punch 2003: 111). This refers to the negligence of a corporation as an organisation or entity, thus rooting liability in 'organisational fault' (ibid.: 110). In defending itself against any prosecution, and to demonstrate due diligence, a corporate body must show that it had considered the possibility of an offence *and* had put into place measures to avoid this outcome. A clear implication is that 'the company needs to have established guidelines and control systems that take into account the risk of the offence being committed. If it has not, then it will be found to be "structurally negligent". 'It is not enough that the company has set up a generic control system' (ibid.: 111). This again is an innovation which attempts to reflect the actual nature and functioning of the modern corporation – and, though not developed for the purpose, there is no reason why it could not apply to safety offences.

In the Netherlands, where there has been a specified principle of corporate liability in the Dutch Penal Code since 1976, the courts have very slowly and tentatively moved towards establishing organisational criteria for attributing liability and fault. The Dutch approach is based upon the two principles of 'power and acceptance' developed in case law. Corporate liability in the Netherlands rests upon showing whether the defendant (the corporation) had the necessary power to determine whether an employee acted in a particular way; and whether the corporation normally accepted such conduct (Field and Jorg 1991: 163–64). The concepts of power and acceptance

> point not to the judgment of isolated positive acts by individuals, but to continuing collective processes. Both suggest more than just a generalised value judgement: ought this to have been allowed to happen? They require concrete examination of the relationship between individuals and corporate groups and a focussing on corporate decision making processes. (ibid.: 166)

This produces, in principle at least, the recognition that corporate crimes are produced by ongoing managerial processes that are routinely tolerated, thus accounting for the fact that, as we note above, corporate crimes often stem from omissions rather than commissions. Yet Field and Jorg's discussion of the development of corporate liability in the Netherlands also raises similar tensions as those discussed above in relation to identification. Although the 'power and acceptance' model allows us to aggregate a series of decisions made (and not made) in the body corporate across time and in a range of different situations, there remains a need to identify where liability rests for the general practices of the corporation. The division of where the power lies in the corporation (or in Anglo-doctrinal terms, the distinction between the 'hands' and the 'brain' of the corporation) is not formally made.

A *de facto* corporate veil

So far, this chapter has argued that the 'identification doctrine' conjures up a rather odd conceptualisation of the relationship between the corporation and the individual. For the purposes of identifying the defendant, the identification doctrine separates the organisation from its constituent parts (its workers and its senior managers and directors). The corporate entity, as in corporate law, is envisaged as a separate person. This corporate person enjoys the same protections and rights as an individual (it has recourse to appeal, to the provisions of the Human Rights Act and so on) but it does not have the same liabilities as individuals. The law insists upon the identification of a rational, free-acting and thinking individual who can be said to possess the necessary *mens rea* before a company can be regarded as guilty.

This has resulted in two key effects. First, reluctance on the part of the CPS to prosecute for corporate manslaughter indicated by the low number of prosecutions taken, and second, the failure of some high profile, and apparently clear-cut, cases. Each of these likely consequences was illustrated starkly in October 2005[9] when Network Rail and Balfour Beatty were fined an unprecedented total of £13.5m for breaches of health and safety regulations related to the Hatfield train crash in 2000[10] (Balfour Beatty's fine was later reduced from £10 million to £7.5 million on appeal). Earlier, Balfour Beatty and five rail executives had faced charges of corporate manslaughter, only for the case to be halted by the judge due to a lack of evidence. Two weeks later, the CPS decided not to proceed with manslaughter charges following the Potters Bar train crash,[11] on the grounds that there was no realistic prospect of conviction; yet in 2004, Network Rail and Jarvis had accepted legal responsibility for claims arising from the crash, to provide 'comfort and assistance' to the victims.

There is no little irony here when we consider the point made in the previous chapter that the criminal law and its guiding principles are designed to deal with individual offending. The criminal justice process takes some odd twists and turns in order to guarantee that individual directors and senior managers are unlikely to be held liable for safety crimes. This is underlined by a clear preference for *corporate* as opposed to individual prosecutions in regulatory cases. Normally, it is the organisation, rather than its senior officials, that is prosecuted for safety crimes. And in most cases, the company faces a fine which, as we will see in Chapter 8, is neither particularly punitive nor likely to secure a deterrent effect. We will also argue in this chapter that the debate on criminalising safety crimes quickly loses its meaning if we only concentrate on the legal process of law itself without considering the form of punishment and what it aims to achieve (Dunford and Ridley 1996; Wells 2001). We will also set out why the use of fines against corporations is not a particularly effective means of punishment. The key point is that corporation can act as a convenient shield for the key decision makers in the corporation. The corporation can effectively absorb the punishment, normally in the form of a fine, while its directors and senior managers are relatively rarely exposed to sanction. Between 1980 and 2004, only 11 directors were convicted for manslaughter. Of those, five were imprisoned, one received a community sentence and five were given suspended sentences. The average custodial sentence was two and a half years.[12] We can say, therefore, that just as the notion of corporate personhood erects a veil which protects both the owners and the senior officers of the company, a corporate veil also exists in a *de facto* sense when it comes to safety crimes.

The UK government's Corporate Manslaughter and Corporate Homicide Bill (see note 4) explicitly denies the possibility of individual liability. Section 16 of the Bill specifically excludes offences that will apply to individuals and notes that: 'an individual cannot be found guilty of aiding, abetting, counseling or procuring an offence of corporate manslaughter'. This clause is particularly interesting since most criminal offences can be prosecuted if someone has been complicit in the crime by aiding and abetting. It is significant in the context of this debate that the Institute of Directors moved from opposition to the Government's reform proposals to vociferous support for a change in corporate manslaughter law (see, for example, *The Safety and Health Practitioner*, December 2002: 4) in 2002, just as the Government's position changed. By contrast, those groups with an interest in the protection of workers and members of the public (such as trade unions and victims organisations) argued equally vociferously for individual liability.

The Bill's rejection of an individual offence contrasts with the position taken by the Canadian government which noted that under their new legislation a corporate executive or board member can also be liable under the Criminal Code for aiding or abetting an offence, counselling a person to be a party to an offence, or being an accessory after the fact to an offence.[13] The Australian Capital Territory (ACT) enacted a separate individual offence in the Crimes (Industrial Manslaughter) Amendment Bill 2003 to allow for the prosecution of 'senior officers' if they negligently or recklessly caused the death of a worker. Senior officers face a maximum penalty of 25 years' imprisonment if they are convicted.[14]

Prosecutions of directors under regulatory law in the UK are also relatively rare. In the ten years up to December 2005, HSC have calculated that, under the HASAW Act, 'proceedings have been brought against a total of 111 directors, of which 86 were convicted and 11 were jailed' (Health and Safety Commission 2005c: 2); moreover, these prosecutions had 'disproportionately impacted on small firms' (Health and Safety Commission 2005d, para 5.2). Similarly low levels of prosecution are revealed in research by the CCA which shows that, in the three years between 2002/03 and 2004/05, there were only 28 convictions of company directors for breaching the HASAW Act.[15] In the context of the total figure of convictions for health and safety offences (around 1,000 a year on average), convictions of directors represent less than 3 per cent of all health and safety prosecutions. We can therefore identify another feature of the *de facto* corporate veil which protects both the owners and the senior officers of the company; a veil that is created in the application of regulatory law.

There are a whole set of separate issues relating to the failure to prosecute directors for HASAW Act offences. The most important reason appears to be that the law does not specify duties for individual directors. This arises from the emphasis in health and safety law upon the duties of 'employers', a term that the courts have consistently interpreted as meaning the company rather than the individual directors of the company. In other words, the duties placed upon companies generally do not impose duties on directors. Section 37 of the HASAW Act does allow for company directors to be prosecuted if an offence has arisen from the 'consent', 'connivance' or 'neglect' of the director. However, the absence of positive safety duties placed upon directors makes the avoidance of prosecution relatively easy because there is no onus on the individual director to take any action to ensure the law is not broken. This relates directly to the points we have already made about acts and omissions in this chapter. It might be relatively easy to prosecute a director for committing a criminal act. But an omission that leads to a crime – which is a much more likely form of culpability for safety crimes – does not necessarily imply a breach of Section 37.[16] The significance of this loophole in failing to provide adequate incentives for directors to maintain a safe workplace is confirmed by a comprehensive review of evidence on the effectiveness of health and safety regulation. This review, *Making Companies Safe: What Works?* (Davis 2004), identifies the need to close the current legal loopholes that enable directors to 'abrogate *all* responsibilities for ensuring that their companies comply with the law and operate safe systems of work' (ibid.: 28; emphasis in the original).

A set of proposals on directors' duties proposed by Glazebrook argues that a tighter form of directors' liability could be secured by the adaption of directors' liability clauses, not only in the HASAW Act, but in a range of regulatory statutes. This would involve the creation of new offences of causing death or serious injury by breaching safety regulations (including the Environmental Protection Act and Nuclear Installations Act, as well as the HASAW Act). The offences would be pinned to directors by asserting that '[e]very person who is (or has the duties of) a director of a corporation is under a duty to prevent the commission or continuance of scheduled offences by the corporation' (Glazebrook 2002: 420). In these proposals, breaches of those offences would carry sentences of up to ten years.

What is rather odd about the current debate about accountability for directors and senior managers is that proposals to improve directors' accountability are generally split between those who advocate the creation of a new crime of violence, and those who would amend existing regulatory law. In the UK, individual accountability is always posed as a choice between the creation of a new crime of violence that can more easily attribute liability for safety crimes to individual directors or senior managers, or amendments to the HASAW Act. In other words, the debate is always framed in terms of a choice between differentiation and assimilation. There is no logical reason not to seek reform by both routes (see, for example, Hall and Johnstone 2005). In most jurisdictions, directors are granted the opportunity in law to hide behind the *de facto* corporate veil that is erected by criminal law. Dealing effectively with safety crimes means that we must adapt the law to deal with the full range of obstacles that lie in the road to accountability for directors and senior managers.

Conclusion

For Glasbeek (2005), the choice facing law reformers seeking to assimilate safety crimes into the mainstream of crimes of violence can be expressed as a choice between corporate personality-based methods of attributing liability (such as corporate *mens rea*) and the creation of a new basis for liability which stretches the boundaries of the identification doctrine. As we have demonstrated, technical legal problems appear to beset both approaches. In sum, the courts have tended to construct an extremely rigid route to the former, and the latter route has tended to lead to a watering down of the criminal status of the offence and the offender. This does not mean, however, that it is beyond our imagination to *develop* laws which seek a new approach to corporate liability which do not decriminalise safety crimes. The route to law reform in Canada, although it also contains some potential flaws, is important because it explicitly attempted to find this balance.

In this and the previous chapter we have discussed the historical and technical processes of 'assimilation' and 'differentiation' in the criminalisation of safety offences. This historical perspective instructs us that there is nothing 'natural' or 'handed down' about the criminal law. Rather, the criminal law evolved in response to a need to maintain a particular social order and to mediate in conflicts that threaten to disrupt that order. The arguments of Carson, Marx, and others[17] are not simply that the criminal law rigidly and reflexively defends the interests of the ruling class to the exclusion of other groups. To maintain order effectively, institutions of the law also need to retain plausible claims to equity and fairness – and on this basis alone, powerful members of society can never be completely exempted from systems of crime control. Thus it is more accurate to see the impulse of Western legal systems not simply as one which protects the powerful, but one which establishes and maintains a particular social order, or, in the words of Carson, a 'viable class society'.

This and the previous chapter have merely provided an overview of the intricate debates and finer points of law and policy that contribute to the predominance of capital in modern liberal democracies. And if this chapter has sought to demonstrate ways in which, and the extent to which, reform in law in this area might be secured, it is to a consideration of regulation that we must now turn in order to further explore these possibilities.

Notes

1 *H.L. Bolton (Engineering) Co. Ltd v. T.J. Graham & Sons* [1957] IQB 159, at 172.
2 *Tesco Supermarkets Ltd. v. Nattrass* ([1972] AC 153), at 171.
3 Those figures are not made public in official data sources. The figures we use are derived from the monitoring work of CCA (see http://www.corporate accountability.org/manslaughter/cases/convictions.htm).
4 The long awaited Corporate Manslaughter Bill was finally introduced in Parliament on the last day of the 2005/06 session. A stated aim of the Bill is to remove the principle of identification, even if this may not be its actual effect, at least in practice; see www.corporateaccountability.org/manslaughter/reformprops/main.htm for a detailed, critical appraisal of the Government's proposals.
5 *Rv. HM Coroner for East Kent ex parte Spooner* (1989) 88 Cr. App. R 10 at 17.
6 In Scotland, where there has never been a successful equivalent corporate homicide offence, the Transco case has effectively put Scottish law on par with the law of England and Wales.
7 Section 12.3 of the Australian Federal Criminal Code (1995). Full text of this section of the code is available at the Australian Attorney General's website: http://scaletext.law.gov.au/html/pasteact/1/686/0/PA000730.htm.
8 Full text of this section of the code is available at the Australian Attorney General's website: http://scaletext.law.gov.au/html/pasteact/1/686/0/PA000740.htm.
9 Source of the following discussion and all quotations is BBC News, *Safety on the Ralways*, at www.news.bbc.co.uk/1/hi/in_depth/uk/2004/safety_on_the_railways/.
10 In October 2000, four people were killed and a further 102 were hurt when a GNER train derailed at Hatfield, Hertfordshire; a broken rail was found to be the 'substantial' cause of the accident.

11 In May 2002, seven people died and 76 were injured when a WAGN service crashed at Potters Bar in Hertfordshire, the triggering cause being a faulty set of points.
12 Figures derived from www.corporateaccountability.org/directors/convictions/manslaughter/main.htm.
13 Full text of the Act can be obtained at: http://www2.parl.gc.ca/HousePublications/Publication.aspx?pub=bill&doc=C-45&parl=37&ses=2&language=E.
14 Full text of the Act can be obtained at: www.corporateaccountability.org/d1/International/australia/ACTManslaughter03.pdf.
15 Available online at www.corporateaccountability.org/directors/convictions/safety/main.htm.
16 For a more detailed analysis, see the CCA briefing at www.corporateaccountability.org/directors/duties/law.htm.
17 For example, Thompson's classic *Whigs and Hunters* (1975).

Chapter 7

Regulating safety crimes

Introduction

A succession of British Ministers have declared that the aim of the Labour Government is to make the UK the most business-friendly environment in the world.[1] Such claims are normally accompanied by the phraseology of 'burdens on business' and 'red tape' to refer to law regulating economic activity, with the unquestioned implication being that such 'burdens' and 'tape' should be reduced as far as possible. In the UK, such terms have entered the political lexicon to such an extent that they are used ubiquitously when the regulation of safety crime, and other aspects of corporate activity, is debated in political circles. As Blair argued in a 2005 speech, this has clear implications for safety regulation and enforcement, 'We cannot guarantee a risk-free life', adding: 'We cannot respond to every accident by trying to guarantee ever more tiny margins of safety. We cannot eliminate risk. We have to live with it, manage it. Sometimes we have to accept: no-one is to blame' (cited *Hazards*, 91, August 2005). To regulate to eliminate risks, he continued, means that 'we lose out in business to India and China, who are prepared to accept the risks'. Notwithstanding that it is no collective 'we' who face risks, but very particular groups of workers (see Chapter 2), this claim also rather obscures the fact that China and India each have occupational fatality rates more than ten times that of Britain (ibid.).

Within this sentiment resides an assumption that to be successful in the modern, globalised world, states have to compete against each other to attract capital. The British government, like most liberal democracies across the world, has accepted the 'realities' of a globalised world order. No matter how contested the idea of globalisation is, there are two major assumptions that are made when the idea of globalisation is accepted as a hegemonic 'truth' or a 'new orthodoxy' (Goldblatt 1997: 140). The first is the assumption that Governments now exert less political control over economies – economic management is relegated to the task of over-seeing the operation of 'free' markets – and over the key actors in these economies, namely corporations and, most significantly, multi- or transnational corporations. The second is that this is an inevitable process, or to recall Margaret Thatcher's words, 'there is no alternative'.

Neither are convincing, since governments still do a great deal of regulating. If we consider that regulation is much more than something state agencies 'do' to business, then the totality of government regulatory activity must be viewed through a much wider lens (Baldwin *et al.* 1998). Governments are continually introducing, refining and enforcing a complex web of regulations that do much more than control harmful corporate activities. Governments set the rules of entry to markets, the rules that allow markets to function efficiently, and establish the infrastructure and the legal framework that allows corporations to employ workers, to produce and exchange goods and services, to invest, speculate and so on. Of course, since this chapter is about safety crime, it specifically addresses the regulation of corporate harm. But at the same time, it is important to bear in mind that safety crime regulation is part of a much larger system of rule making and enforcement.

The reconstruction of government controls on the harms caused by business as 'burdens' on business or as 'unnecessary red tape' is a common theme in contemporary debates on the regulation of safety crimes. It is a theme that is sustained by particular theoretical claims vis-à-vis the nature of corporate offending, the relationship between employing organisations and their employees, and ultimately about how we understand the role of business activity in society. This chapter begins by providing an introduction to the way in which such theoretical claims underpin the approach to regulation adopted in the UK, and examines how they shape the regulation of safety at a workplace level, before turning to explore the range of theoretical explanations that inform our understanding of safety regulation. In essence, this chapter shall explore the origins of consistent under-enforcement that we documented in Chapter 4.

Regulation-in-action: dimensions of under-enforcement

The key, overarching piece of relevant legislation in the UK is the Health and Safety at Work Act 1974 which is underpinned by the philosophy of self-regulation. The Robens Committee, which sat between 1970–1972 to review the existing state of occupational safety (and health) legislation in the UK, concluded that the most fundamental defect of the statutory system was that there was too much law. This resulted in 'apathy' amongst those at work, since 'people are heavily conditioned to think of safety . . .

as . . . a matter of detailed rules imposed by external agencies' (Robens 1972: 7). Moreover, Robens argued that safety is an area in which there is a far greater 'identity of interests' between the 'two sides' of industry than most other aspects of workplace relations. Concluding that the 'primary responsibility' for improving occupational safety lay 'with those who create the risks and those who work with them' (ibid.: 152), the Robens Committee urged the establishment of a 'more effectively self-regulating system' (ibid.). Robens' recommendations concerning the nature of this self-regulating system were formalised in the subsequent HASAW Act 1974, and later provided a model for occupational safety legislation in Canada and a number of Australian states. According to Robens' version of self-regulation, then, criminal law and its enforcement has a fundamental, but minor, role to play in ensuring occupational health and safety protection – the principal responsibility for achieving protection is to be left to those who create and work with the risks, namely employers and employees (see Tombs 1996).

Without entering into detailed discussion of the Act here (see Dawson *et al.* 1988), it is important to highlight key elements of the self-regulatory philosophy, developed by Robens, subsequently institutionalised by the HASAW Act. Regulatory inspectorates, brought together under the umbrella organisation of the HSE, were not conceived of as any kind of 'police force' for industry. Rather, employers and employees should co-operatively self-regulate (Robens 1972: 18–19), through formal and informal structures, while the HSE and its inspectorates disseminate advice, encouraging and overseeing compliance (ibid.: 80). Key elements of this regulatory approach are the use of negotiation and bargaining on the part of regulators to raise, incrementally, general standards of safety management rather than to ensure compliance with detailed, prescriptive regulations. Emphasis is upon prevention, which is represented as distinct from punitive modes of regulation (Pearce and Tombs 1990). Nevertheless, Robens was clear that 'flagrant offences call for the quick and effective application of the law' and that enforcement should be 'rigorous where necessary' (Robens 1972 : 80).

The HASAW Act certainly ushered in some significant improvements to existing safety legislation (Pearce and Tombs 1991: 422), yet in fundamental respects it remained wedded to long-standing assumptions regarding appropriate modes of regulation (Nichols 1990: 330). Critics unsurprisingly rejected its assumption of a fundamental unity of interests within the workplace, and urged the establishment of something akin to a safety policing agency (Nichols and Armstrong 1973; Woolf 1973).

But despite its assumptions of a basic mutuality of interests between employers and employees, even the self-regulatory philosophy of the

HASAW Act 1974 recognised the likelihood of some non-compliance with safety regulations, and the need to *force* compliance upon employers. This pressurising role was assigned mainly to trades unions within workplaces, and to regulators as a source of external pressure. However, this also meant that if either of these should become unable adequately to fulfil their role, then self-regulation could give way to deregulation.

First, then, this philosophy places an enormous onus upon the balances of power – within and beyond workplaces – between capital and labour. At the policy level, self-regulation is linked to a system of tripartism, whereby employees and employers determine policy within the HSC – though this is clearly an organisation in which employers' interests predominate (Dalton 2000). At the level of workplaces, the balance of power is intimately related to the level and strength of the workers' organisation, not least because subsidiary legislation grants formal roles to trade union representatives in the organisation of health and safety. This is a crucial point, since the most well-cited, and widely accepted, finding regarding safety protection at work is that it is best delivered by strong, active trades unions working through safety committees and safety representatives (James and Walters 2005: 98–102) – with HSC finding that workplaces where these conditions are in place experience over 50 per cent fewer injuries than those where such conditions are absent (ibid.: 99).

Second, in terms of the functions of law, it must be clear that any system of self-regulation is predicated upon a range of credible enforcement techniques to which regulators have access and which allow an escalation of sanctions if the regulated body fails to co-operate. Such an enforcement philosophy, however, can only effectively function where three conditions are met: (i) that inspectors actually have a credible presence (to inspect routinely, to investigate incidents) within workplaces; (ii) that escalation towards greater punitiveness on the part of inspectors is possible and, where the circumstances demand it, likely; and (iii) that sanctions formally at the disposal of both regulators and, then, the courts are credible ones. Yet both historically and currently, it is difficult to see any of these conditions of existence of a Robens-style enforcement philosophy actually in place.

Examining the work of HSE inspectors[3]

It is notoriously difficult to gain a detailed insight into what HSE inspectors actually do. Frequency of inspection has always been a thorny issue in relation to the enforcement of safety law. Under-inspection is, as we note in Chapter 4, a primary factor in keeping detection rates low. In its evidence to the Robens Committee, the TUC had 'reminded Robens that the ILO in 1961 had recommended that workplaces should be inspected at least annually' (cited in Dawson *et al.* 1988: 224), a level of inspection that is very far from the original target, with employers being inspected anywhere between once every ten and once every 20 years. As we saw in Chapter 4, low rates of investigations after incidents have been reported to HSE remains another recurrent issue of contention.

Thus the HSE has *never* been granted the resources to act as any kind of police force for industry. At April 1, 1976, after the formal establishment of HSE, there was a total of 3,282 HSC/E staff in post (including temporary staff). This rose to a peak of 4,226 total staff in post in 1979, before beginning an almost year-on-year decline up to 1990, when resources were increased, leading to a new peak of 4,545 staff being reached in 1994. Since that time, numbers have fluctuated, but total staff has never surpassed that 1994 total. At 1 April 2005, there were 3,903 staff in post. Of course, it should be emphasised that, in any one year, a minority of these staff are actually inspectors, the remainder being engaged in a range of support functions. Budget pressures have begun to force resources even lower. A further 250–350 posts are to be lost by 2008, and this, combined with real terms decline in HSE's grant since 2002, means that 'HSE will by 2008 have lost around 17 per cent of the staff it had in 2002 when comparing like with like' (Trades Union Congress 2006).

Just as the only way to relieve the pressure on front-line HSE staff is to reduce their responsibilities and reduce the amount of work they do, the ability of HSE and its inspectorates to do what they are legally charged with is dependent upon the volume and range of regulation to be enforced, the loss or addition of new areas of enforcement responsibilities, and the numbers and types of workplaces[4] across which law is to be enforced. There is therefore a close relationship between the structure and style of the regulatory regime and the resources available to do the job of regulating.

Low rates of investigations *may* indicate a low level of actual offending – so that, for example, the vast majority of occupational injuries do not occur as a result of safety offences. This does, as we argued in Chapter 4, fly in the face of almost all existing available evidence on the scale of safety crimes. Moreover, given the level of discretion of enforcement officers, here as in other areas of criminal justice, the evidence outlined in Chapter 4 is more plausibly a reflection of the general enforcement strategy of HSE (Whyte 2006) and the constraints placed upon it by increasingly scarce resources.

When HSE *do* conduct investigations in the course of routine, proactive inspections, or following incidents that have been reported to them, their primary, often sole, concern is with locating source problems and recommending remedial measures, if they deem any are required, in order to prevent future occurrence of such an event. The HSC's enforcement statement notes that '[g]iving information and advice, issuing improvement or prohibition notices and withdrawal or variation of licences or other authorisations are the main means which inspectors use to achieve the broad aim of dealing with serious risks, securing compliance with health and safety law and preventing harm' (Health and Safety Commission 2004a: 4). Rarely do inspectors seek to gather evidence or information upon which any future prosecution might be based. In other words, the HSE inspection mindset is not one that is geared towards the detection of 'crime' or 'criminals'. Bergman noted that inspectors consistently fail to examine the role of senior company officers in relation to deaths at work, since they 'are not viewed as potential criminals whose conduct requires investigation'(Bergman 1994: 97). More generally, it is widely accepted that inspectors neither enter premises in order to seek out violations, nor respond to the vast majority of observed or known offences by resort to formal enforcement action. Except in the case of the most egregious safety offences, enforcement action is only invoked where processes of persuasion, negotiating and bargaining, often over a very protracted period, have proven 'unsuccessful'. Law is, indeed, the 'last resort' (Hawkins 2002) – a point noted in Chapter 5 and one to which we shall return below.

HSC's recently (2004) launched *Strategy for Workplace Health and Safety in Great Britain to 2010 and Beyond* – a rather bland document of very little substance – further downplays the role of formal enforcement in the work of the HSE inspector, dedicating two paragraphs of its 17 pages to this issue. This is one index of what appears to be a shift in HSE enforcement practice further towards greater voluntarism. Evidence of such a shift is also to be found in a discussion paper by HSE's then Deputy Director-General, Justin McCracken, urging a shift of emphasis to 'educate and influence', so 'using a smaller proportion of [the HSE's] total front line resource for the inspection and enforcement aspects of [HSE's] work'.

Opposed to the general direction in HSE enforcement policy, a July 2004 Parliamentary Select Committee added the recommendation that 'the HSE should not proceed with the proposal to shift resources from inspection and enforcement to fund an increase in education, information and advice', the central component of HSE's new enforcement strategy. 'The evidence supports', it continued, 'that it is inspection, backed by enforcement, that is most effective in motivating duty holders to comply with their responsibilities under health and safety law' (ibid., para 142). Now, despite the Select Committee's findings, and a major meta-review of existing research that concluded that law enforcement is a necessary if not sufficient condition of effective safety protection (Davis 2004), the Government has remained committed to a path away from already meagre levels of enforcement, and towards even greater focus on 'advice' and 'education'.[5] As an internal HSC paper, entitled *Becoming a modern regulator* (dated March 23, 2004) noted, 'there has been deregulatory pressure from within government to reduce burdens on business, be clearer about the benefits of regulation, and more sympathetic to business needs', adding that 'HSE has responded positively to the debate' (cited in *Hazards* 88, October–December 2004). Here there truly are echoes of the response of the first Factory Inspectors to the advice from Government that, in the context of enforcing the 1833 Factory Act, they should co-operate closely with employers so as to make the new law acceptable to them (see Chapter 5).

Subsequently, the Government and HSC launched a complementary set of initiatives around regulation and enforcement. The Hampton Review – established by the Treasury in 2004 to 'consider the scope for reducing administrative burdens on business by promoting a more efficient approach to regulatory inspection and enforcement without reducing regulatory outcomes'[6] – published its report in March 2005, tellingly entitled *Reducing Administrative Burdens: Effective Inspection and Enforcement.*[7] The Hampton Report called for more focused inspections, greater emphasis on advice and education, and in general a removal of the 'burden' of inspection from most premises, and, with full acceptance by the Government, 'were aimed at promoting effective, risk-based enforcement'. Also in March 2005, the Cabinet Office's Better Regulation Task Force published its review of regulation, *Less is More: Reducing Burdens, Improving Outcomes.*[8] The recommendations of these reports came together in the Legislative and Regulatory Reform Bill,[9] which was published in January 2006. The new Bill aims to 'enable delivery of swift and efficient regulatory reform to cut red tape' (Cabinet Office News Release January 11, 2006).

In this fiercely anti-regulation political context, HSC had launched, in July 2005, its own review of regulation, under the rubric of 'a debate on the causes of risk aversion in health and safety', and, in the words of safety minister Lord Hunt, called for 'a common sense approach to risk management'. Thus 'excessive risk aversion does damage too. It hits organisational efficiency, competitiveness, restricts personal freedoms and damages the cause of protecting people from real harm'. Similarly, HSE Deputy Director General Jonathan Rees added that 'HSE's approach to regulation is very much based on sensible risk management. Risk is ubiquitous. Some degree of risk whether financial, environmental or in terms of safety is necessary for progress'.[10]

Unsurprisingly, then, this recent period has seen a renewed interest in experiments aimed at replacing 'traditional' modes of regulation – that is, inspection. Among the experiments in shifts from enforcement that have been rolled out by HSC/E in the past year are the following:[11]

- HSC's intervention strategy, a September 2005 Report, committed HSE to piloting 'ways to identify and recognise good performance', meaning that in exchange for commitments to institute certain safety measures, companies can gain exemption from inspectors' visits;
- HSE's draft 'simplification' plans, published in November 2005, set out HSC/E's 'determination to develop legislation that is easy to understand and comply with to help secure stronger commitment from business', supporting a 'risk-based, targeted approach to enforcement';
- in October 2005, the HSE, in co-ordination with local authorities, launched a Large Organisations Project Pilot (LOPP), which may prove a key, fully-blown self-regulating alternative to enforcement, and which at the pilot stage covered about 1 million workers;
- a Department of Trade and Industry's (DTI) enforcement pilot, begun in June 2005, which uses a 'traffic light' system in which businesses which have demonstrated excellence in health and safety and food standards are no longer subject to inspection.

There is evidence that this shift away from enforcement is beginning to bite. A sharp down-turn in enforcement activity (outlined in Chapter 4) has been accompanied by a reduction in inspections. In the Field Operations Division of HSE, the inspection rate has fallen from an average workplace inspection every 6.99 years to every 10.84 years between 2001/02 and 2004/05. With a reduction in scheduled inspections, and other visible shifts away from enforcement activity, it is inevitable that fewer safety crimes will be brought to the attention of regulatory authorities, never mind the courts. Low rates of prosecution for safety crimes outlined in Chapter 4 can be largely explained by both the acceptance of the importance of industry's commercial 'imperatives' by the HSE (Tombs and Whyte 1998), and its view of itself as a body which must co-operate with and advise industry (Pearce and Tombs 1998). The Hampton Agenda and the associated policy shift further towards a model of self-regulation is driving down levels of prosecution and other enforcement activity. The problem of regulatory under-enforcement is therefore likely to get much worse. The effect is that self-regulation gives way to deregulation, and, as we have seen in Chapter 4, more and more safety offences effectively become decriminalised.

Government resourcing commitments which ensure that there are fewer HSE inspectors than traffic wardens in central London (Trades Union Congress 2006) and concomitant low levels of investigation and inspection, will reduce yet further the pool of incidents upon which inspectors can intervene. Beyond the issue of 'visibility' of safety crimes, the discussion set out here raises major questions about how this shift away from enforcement towards self-regulation, persuasion and bargaining can be legitimised. It is to a consideration for the various theoretical understandings of, and justifications for, regulatory 'under-enforcement' set out in this chapter and in Chapter 4 that we now turn.

How and why is law enforced? Theorising regulation

The picture above – of consistent under-enforcement of safety law and the push for more, rather than less, enforcement activity – is typical. There now exists a mass of studies – mostly nationally based, though with some useful cross-national comparative studies also, some on safety in particular, but many on other forms of regulation of business activity – regarding the practices of a whole range of regulatory bodies. Of the few, broad generalisations that can be drawn from these studies about the practices and effects of regulatory enforcement agencies, non-enforcement of law is the most frequently found characteristic (Snider 1993: 120–4).[12]

Now, within this literature, where the concern is regulatory enforcement *per se*,[13] it is generally recognised, and here this is particularly the case with regard to 'social regulation', that across both business sectors and discrete areas of legislation, one mode of enforcement predominates, namely a compliance-oriented approach. Indeed, some have argued that there is a generalised convergence across enforcement bodies, jurisdictions, bodies of law, and so on, towards such an enforcement approach (Hutter 1997: 243). Within this enforcement mode, as we shall see in the following section, regulators overwhelmingly enforce through persuasion – they advise, educate, bargain, negotiate and reach compromises with the regulated.

If we have demonstrated, to this point, the fact of regulatory under-enforcement, we still have to *explain why*, rather than simply report that, for instance, regulators are so reluctant to prosecute even when detected offences are deemed serious and sufficient evidence of those offences has been gathered (Hawkins 2002). Regulation needs to be theorised rather than simply described; and, as we have indicated in Chapters 5 and 6, part of this theorisation is understanding its emergence in an historical perspective (for example, it was in our discussion of the early Factory Acts that we were able to comprehend the historical basis for a strategy of prosecution as last resort on the part of the factory inspectorate). Indeed, a wide range of explanations for, and ways of thinking about, the phenomenon of 'regulatory under-enforcement' have been developed over the past 150 years or so. We will now consider the most influential perspectives that have been developed in this body of work.

Consensus theories of regulation

First, we wish to consider a variety of perspectives grouped together under the umbrella term 'consensus'. While there are clearly differences between these, they all share a series of fundamental, pluralist assumptions, including the idea that power in modern social orders is dispersed rather than concentrated, that a variety of interests can be mobilised to influence the formal political agenda (which is where effective power lies), and that while the basic contours of these social orders are supported by a general social consensus, social change, through mobilisation of interests, is possible (see Pearce 1976: 38–41). In effect, these are liberal theories of regulation.

One perspective that has been loosely described as the 'public interest' approach (for discussions see French and Phillips 2000; Baldwin and Cave 1999; Ogus 1994) looks at regulation as a process that occurs as a protective state response to the public 'good' being threatened. In this perspective, systems of regulation are the outcome of dialogue between a range of diffuse competing interests. Regulation is conceived of as the outcome of an open and ultimately benevolent decision-making process that reaches a conclusion after weighing up the pros and cons involved in imposing new restrictions upon business. The assumption in public interest models is that a cross-industry consensus involving relevant stakeholders (generally understood as workers, employers and government) is required for a system of regulation to work.

A body of work produced from a consensus perspective has been developed by a group of researchers mainly based in the Oxford School of Socio-legal Studies (Baldwin 1995, 1997; Black 1997; Hawkins 1984, 1997 and 2002; Hawkins and Thomas 1984; Hutter 1988, 1997, 2001). According to this 'compliance school' of regulation, the most successful regulatory strategies are likely to be those involving persuasion, bargaining, and compromise between regulator and regulated. 'Self-regulation'[14] is therefore to be encouraged, since corporations do not respond particularly well to – indeed they are more likely to be alienated by – the full force of the criminal law. From this perspective, corporations have the capacity to act as good corporate citizens, capable of responsible and moral decision-making. Therefore, regulators must appeal to the better nature of corporations by nurturing co-operative relationships with managements. Compliance writers argue that building a consensus around appropriate forms of corporate crime control is a pre-condition of effective regulation. For compliance styles of regulation to work, they must secure the broad agreement and political support of the 'regulatory community'.

When a consensus is achieved and maintained through a close relationship between regulators and regulated, businesses can be expected to self-regulate effectively (Hutter 1988; 1997). To those ends, strict application of the law is inappropriate for corporations (Bardach and Kagan 1982), and it is therefore important from this perspective that regulators are flexible and use discretion in determining which rules to apply (Black 1996; Hawkins 1996a; Lange 1999). Thus, according to Hawkins (1990: 461): '[g]iven the vast numbers of regulatory violations, many of which would be widely regarded as minor, others of which would be seen as serious or even potentially disastrous, and given the scarcity of regulatory resources, a more discriminating approach to regulatory enforcement is needed'. Consensus theories thus support the view that regulatory officials must act as consultants rather than police (Pearce and Tombs 1990).

Now, on one level, the characterisation of the approach to enforcement on the part of inspectors entailed in this perspective is an accurate one in descriptive terms – witness the detailing of such attitudes in Hawkins's recent study of prosecutorial decision-making in HSE (Hawkins 2002). Thus, from this viewpoint, the under-enforcement of safety law that we described in the earlier sections of this chapter is inevitable, since it is a reflection of organisational mission and values, refracted through organisational cultures. However, the compliance school goes further than this, arguing not only is such an approach to enforcement inevitable, but that it is both necessary and desirable. In other words, they move from describing, usually through inspectors' own rationalisations, a situation *as is* to arguing that this is how things *must* be and indeed how they *should* be, slipping between the empirical and the normative. Thus the majority of texts in this area both *document* the predominance of this approach, and *endorse* (some version of) this as the most appropriate enforcement approach.

A variant upon these arguments, but one which shares many of their basic assumptions, is to be found in the voluminous work of John Braithwaite and a series of colleagues over the past 20 years.[15] This influential work, whilst recognising that there are a range of options along the regulatory continuum, argues that 'monitored' or 'enforced' self-regulation is the most pragmatic regulatory outcome (Braithwaite 1982; Braithwaite and Fisse 1987). Braithwaite is in favour of compliance strategies, but only when the threat of tough sanctions remains an option to regulators (Braithwaite 2000). In one of the most sophisticated arguments for self-regulation, Braithwaite and Fisse argue from the starting point that state regulators will never have the resources to enforce regulatory law effectively, whilst adding that internal regulators enjoy certain technical and social advantages over those on the outside (Braithwaite and Fisse 1987). Self-regulation is described and prescribed as being based upon a 'carrot and stick' approach: where self-regulation proves ineffective, the next preferred regulatory tactic is to move to 'enforced self-regulation', this requiring a company to develop a tailored set of rules by which it intends to comply with law which, once approved by external regulators, would then be 'enforced' internally. Where evidence of non-compliance emerges, the potential of punitive external intervention remains (Ayres and Braithwaite 1992: 102–16).[16]

Strategies of enforcement are thus conceived of in terms of a pyramid, where non-compliance leads to the invoking of ever more interventionist or punitive modes of enforcement on the part of regulators. In essence, this is an incrementalist and compliance-oriented strategy based on the principle of deterrence. The issue of deterrence is also considered at length by Braithwaite herein, and elsewhere. In general, Braithwaite argues that there are some contexts in which deterrent-based strategies work, whilst in others these are simply counter-productive. Thus, deterrence must be 'integrated into a strategy that tries persuasion first . . . then incapacitation where deterrence fails' (Braithwaite 2000: 114). Regulatory agencies are best able to do their job when they are 'benign Big Guns' – that is, when they view corporate actors as both rational and moral actors, but assume the latter until proven wrong, thus emphasising persuasion and self-regulation until the point at which greater punitiveness becomes necessary.

Increasingly allied to arguments around self- and enforced self-regulation are descriptions of, or prescriptions regarding, the utility of 'goal-oriented' over prescriptive ('command and control') legislation. While the latter is said to specify the means of securing compliance, the former entails an agency negotiating 'the substantive regulatory goal with industry, leaving the industry discretion and responsibility of how to achieve this goal' (Ayres and Braithwaite 1992: 38). This process of negotiation, based upon a dialogue or conversation both depends upon, but is also one way of achieving, shared, informed interpretations (and the norms, values, and conventions that give rise to them). Thus, 'interpretive' or 'regulatory' communities allow mutual understanding to be secured without further resort to rules.

'Consensus' has been a dominant guiding principle for regulatory policy in the UK, but also in virtually all capitalist economies. Thus, for example the Robens philosophy that, as we have noted earlier in the chapter, provides the foundations for UK Health and Safety regulation (and, indeed, in other jurisdictions), is based upon the idea that there is a natural affinity between senior managers and workers on workplace safety (Nichols and Armstrong 1973; see also Nichols 1990). There is therefore an important link between theory and policy to be observed here. Consensus theories of regulation tend to reflect the dominant themes in 'official' doctrine and policy on controlling safety crime. As a result, we might say that consensus theories analyse the regulatory process from a perspective that incorporates many of the assumptions that governments, regulatory officials and businesses use to inform policy. This conflation between what 'is' and what 'ought' to be renders consensus and compliance theories relatively difficult to criticise (consensus theorists can always deflect criticism by saying they are merely describing rather than prescribing; see Hawkins 1990 for an example of this defence). Despite this tautological insulation from critique, consensus and compliance theories have been exposed to sustained criticism. Five of those main criticisms can be summarised as follows.

First, some commentators argue that there is little empirical evidence for the claims of consensus theories about the desirability of self-regulation (Dawson *et al.* 1988; Smith and Tombs 1995). Indeed, while its desirability is often based upon the undesirable nature and effects of the enforcement approach with which it is contrasted – so called 'punitive' enforcement – the case against strict enforcement regulation is always made hypothetically, since it has never been tried in practice over a sustained period (although see Alvesalo and Tombs 2001; Alvesalo 2003b). At the same time, there is a significant body of *official* evidence that contradicts the feasibility of self-regulation, not least in post-disaster inquiry reports. Thus, for example, Lord Cullen's enquiry into the Ladbroke Grove train crash identified shortcomings in the enforcement of health and safety legislation. Criticisms by then Director General of HSE Jenny Bacon went beyond the issues of HSE resources, and she identified a lack of vigour in the HSE Railway Inspectorate's regulatory duties and the 'placing of too much trust in the duty holders [employers]' (Cullen 2001: para.10.18).

Second, critics argue that consensus models fail to recognise how the concentration of power in business organisations impacts upon strategies of control of safety crime. Although some consensus models take some account of the wider social and economic frame within which regulatory strategies are shaped (for example, Grabosky 1997; Hawkins 2002), this is too easily relegated to one among a number of factors, and certainly the immense resources of financial and political power that (particularly larger) corporations draw upon to bargain with governments and with regulators are glossed over in such models of regulation (see Hawkins 1984: 192).

Third, consensus models see conflicts over regulation as a peripheral rather than central feature of regulation. Corporate or organisational resistance to regulation is precipitated by small numbers of malicious deviants or 'bolshie types' (Hawkins 1996b: 312). Workers and popular movements against corporate crime are also characterised as peripheral, and rarely warrant more than a passing mention in many of those texts. Accounts of regulation from this perspective too easily forget two lessons from history. On one hand (as we saw in the case of the emergence of the Factory Acts) historically, businesses and their representatives have fought bitterly in opposition to regulation when it is not in their clear interest. To this we can add a large contemporary body of evidence that demonstrates how companies obfuscate, lie, cheat and make threats to disinvest, often engaging in fierce public relations campaigns and behind-the-scenes political maneouvres to avoid regulatory reform (Monbiot 2000; Tombs and Whyte 1998; Woolfson *et al.* 1996; Tweedale 2000). On the other hand, consensus models fail to recognise that regulatory controls have often been established only after long and bitter struggles by organised groups of workers and other social movements. A recent example of this tendency is exhibited by Hawkins in his book *Law as Last Resort* (2002). In this text, the only nod to the significance of social movements that support progressive social regulation is found in one reference to Ralph Nader (Hawkins 2002: 14), and even then, this is to set up a crude (and misleading) link between his own criticisms of command and control regulation and Nader's attack on corrupt regulation. The book fails even to make mention of organisations such as Hazards or Disaster Action, or other high profile public/trade union organisations, that have consistently, often very effectively, campaigned for more effective regulatory enforcement. The cases of Simon Jones, Sonae, Bhopal, and the *Herald of Free Enterprise* (as we saw in Chapter 1) are all further testimony to the importance of local, national and indeed international protest movements in pursuing enforcement and/or prosecution in particular cases of safety crimes.

Fourth, because it is myopic when it comes to this aspect of the regulatory dynamic, consensus theorists are highly selective about the 'consensus' they describe. As Davis has argued in the context of the UK, the criminal law has failed to reflect the growing consensus in public support for greater corporate and managerial accountability for deaths at work (Davis 2000: 18).

Fifth, the assumption that corporations are essentially responsible, moral, decision makers underplays both the routine and pervasive nature of corporate offending and the fact that social regulation is often adopted after corporations have committed the least responsible, reckless and most socially damaging acts. Moreover, this assumption obscures the techniques of 'creative compliance' that corporations use systematically to avoid the law (McBarnet and Whelan 1991; McBarnet 1988). Finally, it rather side-steps the *sina qua non* of the corporation – that it is legally bound to maximise profitability for its shareholders.

Neo-liberal theories of regulation

Neo-liberal theorists, most closely associated with the Milton Friedman's Chicago School of Economics, argue that, in general, states have an interventionist tendency that obstructs the efficient regulation of economic activity. Neo-liberals argue that we are over-regulated. Thus, within this general perspective, better safety protection should be secured not through legal regulation and its more adequate enforcement, but through the natural, invisible, disciplining hand of the market.

Market mechanisms are constructed as more appropriate and efficient means of allocating resources and exerting control on participants in markets than governments. Neo-liberal market logic holds that workers enter employment after freely agreeing contractual terms with employers. The risks that workers are exposed to will have a nominal value in those negotiations, and employers will find an optimum level of safety provision that is necessary to attract workers on competitive wages (Moore 1991). Where this optimal level is not reached, then an excess of injuries and the subsequent reaction within the labour market will naturally adjust to eschew employment in a particular firm, or encourage premiums to be paid for 'danger'. This market adjustment results in higher prices than competitors for similar good or services, and thus a loss of business. Thus, the working of the market, left freely to operate, will correct the original 'error'. No state intervention can achieve optimal efficiency so perfectly. However, if neo-liberals are opposed to legal interventions by the state but are broadly supportive of the individual's right to seek legal redress, so, if the market mechanisms fail workers or consumers, then compensation can be provided by the civil law resort to litigation.

The neo-liberal economic perspective is of course associated with the policies of the Thatcher and Reagan governments of the 1980s, but latterly has been embedded in subsequent governments in the US and the UK albeit in a range of slightly modified forms (Jessop 2002). A key element of the neo-liberal ideology that took hold of political systems in the late 1970s and 1980s was the institutionalisation of 'deregulation' as a centre-piece of economic policy. In fact, this was a highly selective deregulation. Whilst states actually expanded their capacity for intervention in social life (for example criminal justice and penal systems in most western states grew rapidly in the same period), some forms of social regulation were deemed unsustainable and damaging to the economy. It was the Reagan and Thatcher governments that first used the language of 'red tape' or 'burdens on business'. In the UK, a Deregulation Unit was established at cabinet level,[17] with the remit to spark a 'bonfire of controls', to take place in concert with a mass privatisation of publicly owned industries. What followed in the US, UK and in other prominent OECD countries was a twin pronged attack upon the legislative safeguards governing some forms of anti-social business activity, perhaps most notably worker safety. This took different forms in different states, depending upon the level of resistance met from pro-regulatory forces (see below). Whether ultimately material or ideological in character, deregulation became the order of the day. And neo-liberal, deregulatory tendencies have been given further impetus as neo-liberalism has, over the past quarter of a century, achieved an international predominance as the 'one best' way of organising economies – a process that is signified by the term globalisation, to which we shall return, below.

Thus an increasingly important arena where we find the influence of neo-liberal ideas is in the global trade regime. Latterly, this ideology has supported the coerced expansion of privatisation and attacks on safety standards and other regulatory protections in the developing world (Michalowski and Kramer 1987; Tombs and Whyte 2003b). Thus, for example, the World Trade Organisation has forced the repeal of food, environmental and workplace safety regulations where they have been deemed to impede 'free trade' (Peet 2003; Wallach and Sforza 1999). Yet it is important to note that despite its seemingly unassailable position in world politics, neo-liberal free market ideology has never been implemented in its purest form. Rather, it has provided a set of guiding principles of minimal interference and enterprise promotion.

The following three criticisms can be made about the neo-liberal perspective. First, neo-liberal explanations of regulation have been criticised roundly for refusing to recognise that capitalist markets – especially those structured along the lines of neo-liberal prescriptions – fail to protect us because they create and support a concentration of power and information in the hands of a small elite (Chomsky 1999; Pearce and Tombs 1998: 17–25; Snider 2003). The need to earn a living, combined with lack of choice and intense competition in job markets, forces many people to work in unsafe workplaces. Second, even the 'freest' of markets do not and cannot operate without the active intervention of states (Jessop 2002; Moran and Wright 1991; Soros 1998: 36–42). As the economic sociologist Karl Polanyi once argued, even nineteenth-century *laissez faire* economic principles had to be enforced by the state (1944/1957: 139). States establish the market conditions, rules and infrastructures within which businesses operate. If we recall the argument in the introduction to this chapter, states do a great deal to 'regulate' industry and commerce – to maintain its steady progress – even where safety crime controls are weak. A third point follows this critique of 'market freedom'. Markets are never likely to be truly 'free' since even the most profit-oriented businesses themselves often recognise that regulation is in their long-term interests. Large firms in particular are generally unwilling to subordinate themselves to the vagaries of the market (Pearce 1976: 82–4).

Even under the most extreme versions of neo-liberal capitalism, then, regulation (even social regulation) cannot disappear altogether – not least for some of the very same reasons that the Factory Acts needed to be ushered in at the birth of fully-fledged industrial capitalism (see Chapter 5). Yet, neo-liberalism has been highly effective in its assault upon workers' safety and pollution standards and in attacks upon the funding of regulatory agencies. On this point, some neo-liberal theorists argue that government regulation, especially in monopoly industries, has a tendency to produce a corrupting influence between large companies and state regulators (Peltzman 1976; Stigler 1971). State regulation will only ever produce unequal and unfair competition in the market place, because it encourages a mutually re-inforcing relationship between governments and big business. In this sense, this body of theorists, sometimes known as the 'law and economics' movement, has similarities to another perspective on regulation: 'capture theory'.

Capture theories of regulation

Capture theories characterise governments and state regulatory agencies as vulnerable to 'capture' by big business. Regulatory capture theory's most well known exponent is Marver Bernstein (1955) who conceptualised the regulatory process as a 'life cycle' where regulatory agencies tended to go through various stages of maturity. In essence, his argument was that regulatory agencies are born out of a general concern about some social problem which appears amenable to a regulatory solution – and, in the early stages, although they tend to be outmanoeuvred by business, regulatory agencies retain a certain zeal (or a strong political will in favour of regulatory control). This zeal ebbs away as the agency reaches maturity, and in 'old age' is debilitated to the point of emasculation in its final 'capture' by industry. Capture is achieved by a mixture of intense corporate lobbying, the consolidation of elite interests in public and private sectors, and a 'revolving door' of personnel between regulator and regulated. The implication of capture theory is therefore that regulation is counterproductive, since it has an inherent tendency to institutionalise corporate influence. Once the capture process has been set in motion, regulatory agencies are doomed to intervene on behalf of the corporations they regulate, and ultimately to advance the interests of big business.

Other advocates of 'capture' depart from life cycle theory and simply argue that in advanced stages of capitalism the state and its administrative apparatuses will become colonised and ultimately controlled by large corporations. This perspective has been used widely to explain the corporate manipulation of government in a so-called 'globalised' world order (see, for example, Sklair 2001). Thus, Noreena Hertz's widely acclaimed book, *Silent Takeover*, argued that corporate power is now unassailable, since 'Governments are now like flies caught in the intricate web of the market' (2001: 140). The influence of unelected, pro-business 'experts' upon UK government (through their role in policy circles around the Prime Minister and the Treasury, or via appointment to the House of Lords), the colonisation of government think tanks by business and the proliferation of government–industry decision-making partnerships all provide us with strong evidence of a 'capture' tendency in contemporary Britain (Monbiot 2000). And this increasingly incestuous relationship between business and government elites has spilled over into the regulatory field.

If one thinks back to the evidence of under-enforcement set out at the beginning of this chapter, we are presented with a powerful case for capture. But the degree to which the government–industry relationship in any capitalist social order is best described in terms of capture remains questionable. For one thing, at the empirical level, many, if not most, of the agencies to which this approach might be applied have hardly had any (early) period of zealous enforcement – the factory inspectorate, for example, almost immediately reached accommodation with the interests of (large) manufacturers and the sympathies of a class-conscious magistracy, as we described in Chapter 5.[18] Further, theoretically, this approach oversimplifies the role of governments as the instruments or pawns of business. The regulatory process is constructed as uni-directional, with decision-making power moving from business down through states and societies. Three points flow from this critique of capture theory. First, it denies or underplays the possibility for resistance, and – like consensus theories – marginalises the centrality of social movements against corporate crime (Kramer 1989). Capture theories view the potential impact of social movements on the regulatory system as minimal since, if, as capture theory argues, regulatory agencies are doomed to lurch from one captured state of existence to the other, there is little point in pro-regulatory movements focussing their efforts on regulatory policy. Second, it cannot explain why at particular moments, 'stricter' regulation that is clearly antithetical to the immediate interests of corporations – but is in the long-term interests of capital as a whole – can be introduced (Marx 1887/1954; Carson 1979; Stolberg and Harris 2003). Third, capture theory can't explain why over time, regulatory agencies can revert to more punitive strategies, even after the point at which they appear to have been captured, or that there may be variations in enforcement policy within agencies (Hutter 2001).

All of this is not to say that the capture thesis does not have any conceptual value, or that 'capture' does not happen to government agencies at particular moments (for example, Carson 1980c; 1982). The value of capture theories is that they are able to highlight systematic bias in regulatory agencies. An alternative approach is to point to the 'balance of social forces' in any regulatory regime. That is, to look at the balance of power between dominant and sub-dominant groups. Few would disagree that in capitalist societies states tend to act in ways that promote dominant business interests above other interests. A second value of capture theories is that they can explain how regulatory regimes can facilitate corporate crimes by establishing the framework for collusion between industry and government officials. In capitalist social orders, the relationship between business and state agencies is bound have a mutual, even symbiotic, political tendency. Indeed, this point is recognised to some extent by each of the perspectives that we have reviewed so far. Consensus theorists accept that regulatory settlements often favour big business. For neo-liberals, over-concentration of power in the hands of business elites is highlighted as a serious threat to market democracy. For capture theorists, the colonisation of the state by business is inevitable in regulated market societies. So, one point that these perspectives all recognise (albeit to varying degrees) is that *state regulation has a tendency to favour organised business interests.*

What is missing from each of those perspectives, however, is a detailed understanding of how regulation is also used by relatively powerless groups to *resist* corporate power and *oppose* business interests. To the extent that it does occur, regulatory 'capture' is a process that is never complete. Regulatory agencies remain vulnerable to external pressures from labour and from public pressure and popular campaigns. This latter point, on the significance of conflict and resistance, is one of the distinguishing characteristics of Marxist, neo-Marxist and radical approaches (grouped together here as 'critical approaches') on regulation and safety crime.

Critical approaches to regulation

Before we examine this latter point, it is worth noting that the capture thesis is often attributed to Marx and to Marxist analyses that over-simplify the capitalist state as an instrument of ruling class domination. Such critiques often trot out the well worn quote: 'The executive of the modern state is but a committee for managing the affairs of the whole bourgeoisie'[19] (Marx and Engels 1968/1890: 37). This quote is usually deployed not to examine why or on what basis Marx might have developed such a view of the state, but – reproduced devoid of context[20] – to demonstrate the extent to which Marx held to a rudimentary or primitive understanding of the state. Our earlier discussion in Chapter 5 of Marx's analysis of the Factory Acts passages noted a subtle understanding of the fragmented and often contradictory way in which regulation develops. In stark contrast to his intellectual reputation as an 'instrumentalist' or 'economic deterninist', Marx portrays the British state as contradictory and institutionally divided over the struggle for regulation. Regulation is passed in the midst of bitter struggles against legal intervention by the factory owners themselves, and capital at some points is itself divided over its opposition to regulation. Although the significance of Marx's analysis for thinking about the complexity of the capitalist state – and in particular the importance of his observations for understanding the regulatory process –

has rarely been acknowledged, it is his insistence on viewing regulation and its enforcement as an outcome of the struggle between antagonistic class interests that has endured in critical and neo-Marxist analyses of regulation. A number of writers have drawn upon this tradition of viewing struggles and conflict between economic groups or social classes as central to understanding the regulatory process (Carson 1979; Davis 2000: 14–18). Mahon (1979) draws upon the work of the Greek Marxist Nicos Poulantzas, arguing that the state is an 'unequal structure of representation' which absorbs and dissipates conflicts between opposing interests. This has the effect of maintaining the long-term stability of the social order. Regulatory agencies are created following particularly contentious conflicts of interests precisely because conflict needs to be absorbed and dissipated. In turn, regulatory agencies ensure that particular business interests are subordinated to the long-term interests of capital as a whole. This is why we might expect tough regulatory responses after a prolonged campaign against, or public debate over, an issue of corporate crime. It is also why we might expect to see regulation implemented against the wishes of business at particular times (Wilson 1980; Woolfson *et al.* 1996). At times some bids from pro-regulatory forces are irresistible. For example, the Offshore Safety Division (OSD) of HSE was established as a new regulator for the offshore oil industry under the rabidly pro-business Conservative governments of the 1990s. The OSD was formed after the moment of political exposure precipitated by the Piper Alpha disaster and the subsequent campaigns of industrial action for safety rights for workers on North Sea oil platforms (Woolfson *et al.* 1996). This reinforces the central point of the critical and neo-Marxist thinkers, then, that regulatory controls can be most effectively fought for at times when the state – indeed the social order – is vulnerable to pressure from subordinate groups.

Consistent with this analysis, some academic writers have highlighted the centrality of 'pro-regulatory forces' in securing demands for tighter regulation, and in ensuring the effective enforcement of the law. Thus, for Snider for example, regulation is best understood as a dialectical process, determined by the outcome of struggle within two sets of relationships, namely the relationship between states and capitals, and the relationship between states and their broad electorates. Thus key factors in identifying regulatory outcomes are: the interests and strength of various forces within capital, the nature and strength of various pro-regulatory groups, and the interests within and strength of local and national states (Snider 1991). This indicates that understanding the level of regulation at any one time is both an empirical and theoretical task.

Such an approach coheres with a wider body of neo-Marxist and critical literature that views the law as a site of struggles over the definition of 'crime' and over who constitutes the legitimate objects of crime control.[21]

Thus, the emergence of the UK Corporate Manslaughter and Corporate Homicide Bill was only possible due to the combined efforts of workers groups, victims and relatives campaigns, supported by the work of critical academics and lawyers. Although, as we indicated in Chapter 6, the campaign for an effective corporate killing law has been by no means 'successful', what this story does indicate is that struggles for stricter regulation can make progress – and gain widespread popular support – even in the face of hostile economic and political conditions. Indeed, the watered down form that the Bill has taken suggests a process of government mediation between pro-regulatory forces and corporations and their representatives, albeit one which has taken unequal account of the competing demands of these groups.

This view of the state as an unequal structure of representation therefore allows us to see how spaces for challenging power are not necessarily closed down or captured, no matter the intensity of a prevailing pro-business ideology. To return to the point raised at the start of this chapter, despite the current rhetoric of neo-liberal discourse, a diminution of safety standards is not a necessary or inevitable consequence of global competition. States as regulators remain dominant players in, indeed essential to, the functioning of the global economy (Jessop 2002).

Conclusion

Of the four perspectives on regulation that we have explored here, it is the consensus perspective – underpinned by the idea that effective regulation is best secured by building a consensus around the regulatory community – that remains dominant in Western regulatory systems. But we should not necessarily think that the perspectives that underpin regulation are diametrically opposed, or operate in a mutually exclusive way. It appears that in the case of health and safety, the rise of neo-liberalism as an organising political ideology in Western democracies remains unchallenged by the consensus approach. As Hawkins has noted, via the idea of the 'surround' within which regulation proceeds (Hawkins 2002: 48–9, and *passim*); macro economic and political factors play an important role in shaping inspectors' world-views. This throws light on the supine response of HSE to the Government's deregulatory messages, and also indicates that inspectors increasingly enter their bargaining and negotiation with employers with some sense of the greater power of capital, the increased urgency of threats to relocate investment abroad, or the realities of diminished state capacities under conditions of deregulation and privatisation, and so on. Under neo-liberal conditions, in other words, punitive enforcement becomes less feasible, and co-operative or compliance-oriented approaches become much more likely. And it is here that we can see – again – a coherence with policy, since the current and recent British Governments are, for all their deregulatory rhetoric, probably less interested in the removal of law *per se* (which, as Thatcher found in the 1980s, invites political confrontation), but much more interested in changing the terms of that enforcement, towards greater compliance-type techniques which imply less actual impact upon business.[22] The prospects for worker protection are not good if this analysis is accurate. If consensus theories have accommodated and provided intellectual legitimacy to the 'deregulation' strategies promoted by neo-liberals, then capture theories cohere with a different set of neo-liberal claims. Capture theorists, on the left and the right of the spectrum, share the idea that in the current social order, there is no alternative to the triumph of the market over state regulation. TINA (there is no alternative) is essentially the normative equivalent of capture theory's resignation to the corruption of state regulation by the market. Capture theory is therefore unable to produce an alternative to the domination of the regulatory system by corporate interests.

Our discussion in this chapter, however, raises the possibility that an analysis of struggles for regulation is of a complexity that moves us beyond a bilinear struggle between state and capital. The complex struggle for regulation is not resolved only in the formal political arenas of government. Neither is it resolved at the level of the regulatory agency or even the individual workplace. As we argued in Chapter 5, the rule of law in capitalist societies is as much about social order maintenance as it is about control efforts *per se*. Crucially, states continue to set the rules of the market nationally and trans-nationally (Underhill 1994) and reconstruct regulatory and infrastructure systems when markets break down (Alvesalo and Tombs 2001). In other words there is clearly room for manoeuvre on the part of states as to how to regulate and enforce – requiring us again to confront the specificities of balances of power within concrete social orders. If both capture and consensus theories are compatible, in different ways, with the current political mantra of 'there is no alternative', then critical approaches open up considerations of precisely what alternatives do exist – and how they might be pursued and secured.

Notes

1 See, for example, Osler 2002, *passim*.

2 The Safety Representatives and Safety Committee (SRSC) Regulations 1978, and tellingly the one part of health and safety legislation of this period that caused real political controversy (Walters 1987: 48).

3 This chapter does not focus upon the work of Local Authority inspectors, known as Environmental Health Officers (EHOS). But it should be noted that these have significant worker safety enforcement functions in almost 1.2 million establishments.
4 For example, since 1974, the numbers, and rapid turnover, of small businesses in the UK has increased exponentially.
5 See, for example, the Government's Response to the Select Committee Report, in the form of a letter from the Secretary of State for Work and Pensions, published as an Appendix to the House of Commons Work and Pensions Committee. *Government Response to the Committee's Fourth Report into the Work of the Health and Safety Commission and Executive. HC 1137.* London: the Stationery Office; and Health and Safety Commission 2004b.
6 Hampton (2005) *Reducing Administrative Burdens: Effectice Inspection and Enforcement.* London: HM Treasury/HMSO, available at www.hm-treasury.gov.uk/hampton, 1.
7 Available at www.hm-treasury.gov.uk/budget/budget_05/press_notices/bud_bud05_presshampton.cfm.
8 Available at www.cabinetoffice.gov.uk/regulation/documents/pdf/br_act_review.pdf.
9 www.publications.parliament.uk/pa/cm200506/cmbills/111/en/06111x--.htm.
10 Health and Safety Executive Press Release E094:05 July 13, 2005, at www.hse.gov.uk/press/2005/e05094.htm. Quotations from this source.
11 Source: *Risks*, no 235, December 1, 2005, p5..
12 While this holds as a generalisation, it remains a generalisation; there are important national differences in enforcement strategies across differences spheres of regulatory activity (Snider 1991).
13 As opposed, for example, to studies of regulatory policy formation.
14 A model of regulation where corporations are trusted to monitor and control their own compliance with the law under a minimalist regulatory framework.
15 See Braithwaite 2000 for an excellent overview of this work.
16 Gunningham and Johnstone (1999) have recently developed this model in order to argue for a 'twin track' approach to regulation.
17 The fore-runner of what is now known as the Better Regulation Executive, referred to earlier in the chapter as leading the review of regulation which resulted in the Legislative and Regulatory Reform Bill (2006).
18 By contrast, some have argued that in the US, the Occupational Safety and Health Administration did begin life with a zealous approach to enforcement, even if this was more expressive than symbolic (Calavita 1983), so that here a capture theory based upon a life-cycle approach might have somewhat more force.
19 Marx's term for the ruling class of property owners.
20 The statement was made in his polemical political programme, *Manifesto of the Communist Party.*
21 The same issues are at stake in campaigns to hold war criminals to account at the International Criminal Court, for the criminalisation of violence against women and in campaigns for the decriminalisation of drug use.
22 In the UK now, any proposal for law reform which may affect business must be accompanied by a 'Regulatory Impact Assessment' which sets out in detail – and costs – the potential impacts that the new law and its enforcement may have upon business.

Chapter 8

Punishing safety crimes

Introduction

This chapter is concerned with how the perpetrators of safety crimes are dealt with after they have been found guilty in a criminal court. Before starting our discussion of how court sanctions and punishment are applied in the context of safety crimes, we begin with a couple of important background issues that need to be borne in mind when reading this chapter. First, it is likely that in most jurisdictions, safety crimes do not attract the most punitive response from the state when compared with other forms of white-collar and corporate crime. Bernie Ebbers, the World Com Chief, was sentenced to 25 years in jail in July 2005 for his part in the corruption that brought down his business empire; and, in October 2006, Jeffrey Skilling, the Enron Chief Executive, was sentenced to 24 years in jail for his part in those frauds. No employer to our knowledge has ever faced such a lengthy sentence for killing – let alone injuring – in the workplace. This can be partly explained as a function of the differentiated way in which states respond to different forms of crime. Large scale crimes of corruption subvert and threaten the smooth operation of legitimate markets in the way that safety crimes generally don't. Therefore it is the former types of crime that may produce more punitive state responses than other forms of white collar and corporate crime (Slapper and Tombs 1999), although as we indicated in the previous chapter, safety crimes and the popular opposition that they often generate can present a crisis of stability for the social order. It is at those moments that law reformers, legislatures and courts are more receptive to arguments for more punitive responses.

The second thing to bear in mind is that this book is concerned with safety *crimes*, and therefore we include no commentary or analysis in relation to the role of civil proceedings in the punishment of corporations. A broad distinction between civil and criminal procedure can be made as follows. The former normally involves an individual initiating a case against another individual or a private corporation to seek compensation. The latter involves an infraction of the criminal law normally (although not always) initiated by a representative of the state. The difference in the gravity or seriousness with which each legal procedure is generally regarded can be illustrated by examining the technical characteristics of each procedure. Criminal courts make a decision on the guilt or otherwise of the defendant; if guilt is proven, the court will normally impose criminal penalties. Civil cases deal with the allocation of costs and damages to one part or the other and normally involve a decision about whether or not the respondent must pay damages. In most civil cases, the plaintiff or applicant seeks damages for a personal injury against them (which may include disease, impairment of physical or mental condition, or death, where the action is initiated by the bereaved). Very few damages cases actually ever reach the courts since the vast majority are subject to out-of-court settlements (Harpwood 1993: 18.1). Personal injury actions against corporations are notoriously difficult to win since corporations have the resources to delay proceedings; there are considerable risks for the plaintiff, since potentially huge court costs have to be paid for by the losing party, costs which are usually much more easily absorbed by the defendant, certainly if the defendant is a large corporation. These factors combine to encourage plaintiffs to settle out of court for a reduced compensation sum rather than seeking full compensation in the courts (Harris 1984). In criminal cases, the relationship between the two parties is entirely different. The state punishes crime on behalf of us all, whereas individuals pursue civil action on behalf of themselves. The fact that punishments can be imposed in the criminal courts once an accused is found guilty further distinguishes criminal procedure from civil procedure. No matter how punitive the latter may often appear, it is criminal cases that imply the state's right to punish, and by extension, the seriousness with which the state regards criminal behaviour.

This point immediately alerts us to how the criminal law acts as a mechanism of social censure (Sumner 1990), or as Mathiesen (1990) put it, a 'message from the state'. The criminalisation of conduct by the State carries a gravitas and social opprobrium that is not comparable with the cases that are pursued in the civil courts simply because the State, rather than the individual, acts as the accuser and the prosecutor on behalf of its citizens.

The argument running through this book is that not all crimes are treated by the state equally; as a result, the social opprobrium carried by legal censure varies across different criminal offences and changes over time and place. Discussions in previous chapters have detailed how safety crimes came to represent a different class of offence. On one hand, safety crimes are very often segregated in law as 'regulatory' or 'administrative' offences, and on the other hand, they tend to be exposed to 'compliance' rather than 'strict enforcement' styles of policing. Moreover, as Chapters 5, 6 and 7 have indicated, the seriousness with which safety crimes are treated is not fixed. This chapter explores how safety crimes are currently treated, and how they might be treated differently, by the courts. It starts by examining how safety crimes fit with the key theoretical perspectives that have been developed to justify contemporary forms of punishment, before going on to describe how safety offences are dealt with in the UK. The final section of this chapter will then propose a range of alternative penalties and sanctions for safety crimes.

Theories of punishment and safety crimes

The observation that deterrence theory is much more applicable to safety crimes when compared with 'mainstream' crimes has become commonplace in the study of white-collar and corporate crime (Sutherland 1983; Chambliss 1967; Geis 1996; Pearce and Tombs 1998). Deterrence theory is based upon the idea that individual conduct – deciding, for example, whether or not to commit a criminal act – is shaped by the costs and benefits that might arise from the conduct. Individuals therefore make a rational calculation that weighs the chances of being caught and the severity of the punishment against the 'benefits' of committing a crime. It is a perspective that is neatly summed up in Jeremy Bentham's idea of *homo economicus*, the self-interested, rational thinking 'economic man'. In theoretical terms, deterrence theory has been most commonly challenged on two counts. First, rational choice depends upon the subject having perfect knowledge of the risks of being caught. Second, rational choice depends upon individuals' being capable of exercising rational judgement. Generally, the model is applied to those who are *least* capable of acting rationally. Pierre Bourdieu (1998: 83), referring to the capacity of lower-status populations to mobilise and improve their conditions of existence, has made this point:

> The unemployed and the casualised workers, having suffered a blow to their capacity to project themselves into the future, which is the precondition for all so-called rational conducts, starting with economic calculation . . . are scarcely capable of being mobilised … in other words, a reasoned ambition to transform the present by reference to the projected future, one needs some grasp on the present.

The ability to act rationally, in other words, is severely compromised where people do not have any control over the social conditions that shape their present and their future.

Conversely, companies and their senior officers do have some motivation to consider the long term consequences of their decisions, and the costs of punishment to their business and their social position. They are much more likely to commit crime only after making a reasoned assessment and choice to act rationally. Corporations are 'future oriented' (Braithwaite 1989). Moreover, although most individuals do not possess the information necessary to calculate rationally the probability of detection and punishment, large bureaucratic organisations do have the resources to deploy sophisticated information gathering systems and to call upon lawyers and accountants. Both companies and their directors do make calculated decisions, not based upon perfect knowledge, but based upon a range of knowledge resources available to them which allow them to make calculated decisions. Chambliss (1967) argued that deterrence can appropriately be used against white collar offenders because they satisfy two conditions. First, they do not have a commitment to crime as a way of life; and second, their offences are instrumental rather than expressive. In other words, the crimes they commit are less likely to be spontaneous or emotional as we might find in many 'street' crimes, but are often the result of calculated risks taken in boardrooms. Organisations and their senior officers use cost-benefit analysis to assess the implications of strategic decisions as a routine procedure.

There are some key examples here. Perhaps the most infamous case is the decision to proceed with the production of the Ford Pinto, despite corporate knowledge of a lethal design flaw, based on the calculation that the explosions that would follow impact upon the fuel tank would cause a projected total of 180 deaths and 180 serious injuries. On the company's reckoning, each death would cost them $200,000, each injury $67,000 and each car $700. Thus while the total cost of allowing the dangerous car to leave the production line was $49.5m, the costs of design change were estimated at $137m; the company therefore decided to produce and market the car with the fault. Dowie's investigation into the Ford Pinto revealed that in fact 500 people had died as a result of Pinto explosions (Dowie 1977). This type of calculation is also used when firms consider major safety investments. The British Railways Board and Railtrack, used a cost-benefit calculation to decide against installing the Automatic Train Protection (ATP) system. The company's estimation was that the system would cost approximately £14m per human life saved. The cost of instalment greatly exceeded the projected economic value of the lives that would be saved (£3m per life). The calculation was brought into sharp relief by the Ladbroke Grove train crash in 1999. Had ATP been installed, it would most likely have saved the lives of 31 people killed at Ladbroke Grove (Whyte 1999b).

Those are perhaps the most dramatic and extreme examples of the application of rational decision-making in business contexts and it is important to note that we would not expect to see this level of analysis in the everyday decisions that businesses make. However, it remains the case that businesses do routinely make decisions based upon the projected cost implications of a particular activity or strategy. In other words, perpetrators of safety crimes are better equipped and more likely to be rationally focused upon the future consequences of their policies, strategies and the way that they organise their activities.

Deterrence theory is the bedrock of contemporary western systems of criminal punishment. However, saying that deterrence theory underpins our approach to punishment is not to say that it is the only philosophical basis that we find influential in systems of punishment. Deterrence theory is typically contrasted with retribution or 'just deserts' theory, the latter conceiving of punishment as a means of repaying society for transgressing rules. The basis for this philosophy is not that citizens do act rationally, but that they *should* act rationally. It is also underpinned by the liberal assumption that the State's rules are there to benefit all. If some choose to break the rules, it is normal to achieve some kind of pecuniary advantage over others. Therefore, punishment should be conceived of as repaying a debt to society. Penal theorists like Andrew von Hirsch (1976/1996) also argue that the just deserts perspective clarifies the appropriate penalty for the crime and therefore improves transparency and accountablility in the penal system. Murphy (1973/1995) has criticised just deserts from a similar position to that outlined above, namely that the retributive conception of justice masks an inherent contradiction: because repaying a debt to society means the least to those who are the least well positioned within the social structure, punishment which seeks retribution is likely to be least effective when applied to the social groups which are, overwhelmingly, the principal target of contemporary criminal justice systems.

Rehabilitative theories of punishment are based upon the idea that punishment should make interventions that are likely to transform the criminal into a law-abiding citizen. Rehabilitation aims to reform the criminal and re-integrate them into the community. John Braithwaite has developed perhaps the most influential perspective on rehabilitation over the past 20 years or so. He argues that rehabilitation fails to work because often punishments are disconnected from the social environment that they seek to change. In order to reconnect the offender to the community that he or she has offended against, Braithwaite's (1989) 'reintegrative shaming' thesis proposes that punishment is less likely to work by exclusionary or stigmatising practices, than if the offender has a chance of being re-integrated into society. Influenced by forms of punishment found in Japanese and in Maori culture, Braithwaite argues that a process of public shaming can facilitate the re-integration of the offender into the community. The criticism that has been levelled against reintegrative shaming is that it is difficult to re-integrate individuals of low social status, since they are likely to be in a social position that is disconnected from the 'community' dealing out the punishment in any case and will therefore not necessarily respond to a process of shaming. However, what is interesting for the purposes of the argument here is that for Fisse and Braithwaite (1983) this principle of re-integration can be most effective when dealing with organisations and individuals of high status in the community, since they have a great deal to lose in terms of reputation.[1]

A final justification for punishment that is commonly applied to western criminal justice systems is incapacitation (Greenwood 1983). In the case of the most dangerous or persistent offenders, prison or intensive monitoring can be used to remove an individual from the social environment he or she is deemed to be a danger to. Incapacitation is therefore normally achieved by long prison sentences. Aside from the brutalising effect that long prison sentences have on those that are subjected to them, the key problem with this approach is that it disconnects the criminal conduct from the motivations or conditions that give rise to the crime in the first place and instead superimposes upon the offender another set of harsh and brutalising conditions. Incapacitation in the context of safety crimes might take a different form. Incapacitation of organisations can be achieved by isolating a particular activity or part of the organisation where offending is identified, or by targeting the whole organisation and intervening to physically prevent it from committing criminal activities. A fuller discussion of the ways in which this can be done is provided below. Incapacitation of senior managers or directors who are found guilty of safety crimes can be achieved by prison of course, but this is problematic because it does not deal with motivation or causation, and, in capitalist social orders, only a tiny minority of criminal directors and senior managers are ever likely to be given prison sentences. We might propose a more viable form of incapacitation that does not aim to remove the offender from society altogether, but only from the community he or she is a danger to. By disqualifying criminal directors and senior managers from holding high office in an organisation, we can ensure that they will be incapacitated and at the same time retain a clear connection between the punishment and the specific conditions that give rise to the crime. It is important not to lose sight of the fact that effective incapacitation might be achieved by withdrawing the privilege to make profits in the case of the corporation, or the privilege to hold high office in the case of an individual who is the target of the sanction. Incapacitation in this context is therefore better understood as incapacitation of privilege (since it is from a position of privilege that safety crimes are committed), rather than isolation from the social body.

In short, then, there is reason to expect that each of the key philosophies of punishment discussed above can be appropriately applied in the use of sanctions against companies and high status individuals. The theoretical bases of each of those philosophies of punishment are challenged when we take into account the social status and relative position of privilege/deprivation held by those that are punished. Deterrence is compromised because lower status individuals are least well placed to exercise rational choice. Retribution is compromised because the concept of the social contract means little to those that have little to gain from fulfilling their part of the social contract in the first place. Rehabilitation is compromised by the false hope of self-improvement when re-integration into society may mean little more than re-integration into the conditions that led one to commit criminal acts in the first place. Incapacitation is compromised because it seeks merely to disconnect offenders temporarily from the social body. When it comes to dealing with safety crimes committed by relatively powerful offenders, however, the key problems in each of those cases might be negated by the relative privilege or relatively higher social status of the offender, particularly if the form of punishment can be shaped in a way that takes account of this position of power and privilege.

In theory, all of this appears relatively unproblematic. But we should be very careful about oversimplifying systems of punishment. It is not enough to say in some general way 'criminal sanctions work', because the effectiveness of the punishment is likely to vary so much depending upon the receptiveness of those punished, the circumstances that led to the offence, the likelihood of future prevention and so on. We therefore need greater theoretical clarity that allows us to see which types of punishments might be most effective when applied to specific criminal offences and when applied against specific perpetrators. In the case of sanctions that seek deterrence, the aim is to create either a general or individual effect of discouraging crime by making the costs too high. In the case of retributive punishment, the aim is to ensure that the individual repays a debt to society that is commensurate with the offence. In the case of rehabilitation, the aim is to reform the individual. In the case of incapacitation, the aim is physically to prevent the perpetrator from committing crime again. It is the degree to which those aims of penal theory might be met that the rest of this chapter seeks to unravel by considering, first of all, how safety crimes are currently punished, and second, how this system of punishment might be reformed.

Fines and safety crimes

Most offences that are prosecuted, including the most serious crimes, will, if the prosecution is successful, incur a fine. Fines following health and safety convictions are notoriously low. This is partly because, although the higher courts are permitted to impose an unlimited fine, the maximum that can be imposed by the Magistrates Courts is £20,000. Since most safety crimes in England and Wales are prosecuted in the Magistrates Courts, the average fine remains low. Although, as Figure 4 below shows, there has been a sharp trend upwards in the size of fines imposed for safety crimes over the past eight years, the average remains under £14,000. The UK Treasury's review of enforcement (Hampton 2005) concluded that fines in the Magistrates Courts were often too low to eliminate the economic benefits derived from law breaking.

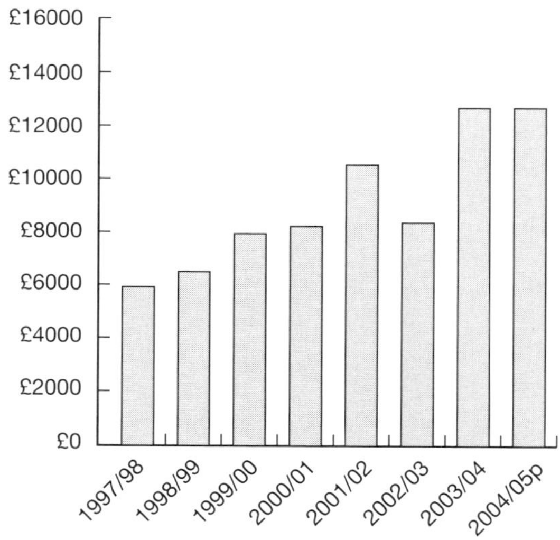

Figure 4 Average fine following conviction of a safety offence in the UK[2]

Figure 4 shows the average fine in convictions of cases taken by the Field Operations Division (FOD) of HSE (which comprises around 60 per cent of total convictions). It shows a steady rise in level of fines imposed by the courts. Another way that this data can be represented is by taking away some of the larger fines that tend to skew the figures. Thus, HSE has published a figure that excludes all fines over £100,000 (which added up to only 13 convictions in 2005/06). The average for all HSE health and safety convictions for 2005/06 when removed of those 13 convictions was £6,219.

As we might expect, the fines for offences that result in a fatality are higher. This partly reflects the gravity of the offence, but it also reflects that fact that those offences are more than twice as likely to be heard in a higher court (Unison/Centre for Corporate Accountability 2002: 15). Figure 5 below sets out the level of average fine for a prosecution following a fatality.

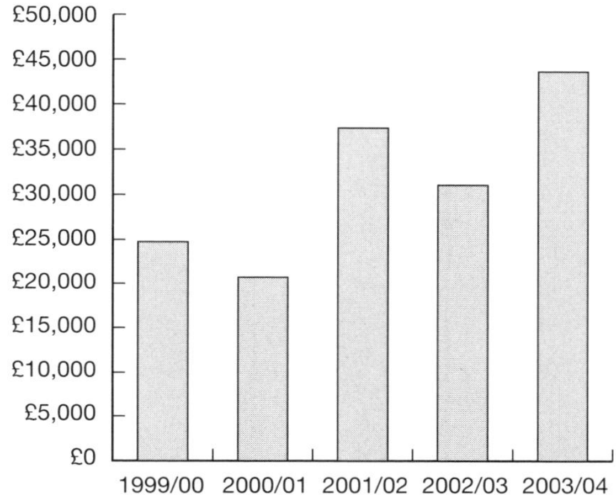

Figure 5 Average fines following work related fatalities in the UK

As this figure demonstrates, the average fine for a fatality has risen steadily in recent years. The first issue that we should note in relation to this data is that it tells us very little about the real impact of fines upon the offender. We know from the preceding discussion that most of the convicted parties are companies, but this is about all that we do know. We have no way of knowing, for example, what those fines are in relation to the turnover or the profits of the firm that has been fined. Of course, a fine of £45,000 is unlikely to make an impact on a large trans-national firm, when compared with, say, a small family partnership. Having said this, we can identify at least five high profile cases involving large firms since 1999 that have attracted fines of over £1m. In 1999, Balfour Beatty was fined £1.2m for the collapse of a tunnel during the construction of a new rail link to Heathrow airport (a record fine for an incident that didn't involve a death). Also in 1999, Great Western Trains were fined £1.5m for causing the Southall train crash that killed seven and injured 150. In 2002, BP were fined £1m for causing an explosion at its Grangemouth plant. In

2005, the gas company Transco was fined £15m for killing a family of four in a gas explosion (see Chapter 1). In 2006, Balfour Beatty was again fined a record-breaking amount. The company was ordered to pay £10m for the causation of four deaths and 102 injuries in the Hatfield crash (later cut to £7.5m on appeal).

Fines of this magnitude, when considered alongside the upward trend indicated in Figures 4 and 5 above, appear on the face of things to indicate a gradual acceptance by the courts that higher penalties should be imposed for safety offences. A system that is almost exclusively reliant upon a system of fines, however, is likely to remain problematic for five main reasons.

First, if we take the large fines noted above, they are very often not particularly large when taken in comparison with the profits of the companies they seek to penalise. They are even less significant when set against the annual turnover of a company – turnover being the appropriate figure if one wants, for example, to measure the impact of any particular fine when compared to a fine against an individual based on his/her annual income. As we have noted already in Chapter 1, the £15m record fine for a health and safety offence in the UK levied on Transco Plc in August 2005 amounted to less than 2 per cent of the previous year's after-tax profit. Another way of expressing the value of the fine is as 0.16 per cent of turnover for 2004, equating to a fine of £40 for someone earning £25,000 a year. This record fine for such a grave offence hardly dented the company's revenue. In relative terms, it is typically lower than that imposed by local authorities upon local citizens who allow their dogs to foul public spaces. And, of course, the problem of the low level of even 'record' fines is exacerbated when the chances of prosecution, as we have already seen in previous chapters, are rather low. Thus, it might be said that a fine even at this record level, might have very little impact upon the management of an organisation unless it knows that either the chances of being caught, or the size of the fine, amount to too great a risk to take. Even the highest levels at which fines are imposed at the moment are hardly likely to provide a deterrent to offending, or to put it another way, can hardly provide an incentive to be law abiding.

Second, because of the tendency for the average fine to be relatively low, fines certainly have most acute effect on the smallest firms. It is small firms that, as we have seen in Chapter 6, are most readily prosecuted for the offence of corporate manslaughter. Therefore the current system of punishment for safety crimes can be said to be inequitably applied to organisations depending upon their size.

Third, because safety crimes are constructed using the 'inchoate mode' (see Chapter 5), fines are primarily related to the gravity of the breach of health and safety/corporate manslaughter law, as opposed to the outcome of that breach. In case law, it has been established that the courts should also consider higher penalties following a health and safety breach if the outcome has been a serious injury, ill heath, or a death, but this is very often a secondary consideration of the courts.[4] This means that fines do not necessarily correspond to the harm caused and may appear derisory when they are imposed to punish debilitating injuries or deaths. In the Transco case, the size of the fine was ostensibly not related to the fact that the firm had killed a family of four in their own home, rather the presiding Judge Lord Calloway noted that fine was imposed on the company because of the seriousness of the offence, and because of the lack of remorse shown by the firm's representatives (a justification that is consistent with *R v Howe* – see footnote 4 – since prompt admission of responsibility can be considered a mitigating factor in sentencing such cases in the UK).

Fourth, because fines are levied on the organisation generally, rather than targeted at a particular group within the company, those costs can be absorbed by the organisation as it sees fit. The costs of even the largest fine might be offset against a particular budget heading (they might result in cuts to running or maintenance costs that may even worsen the management of safety in an organisation), or they may be passed on to customers and clients in the form of price rises, or to suppliers by reducing the market value of a product. The costs of fines may even be passed on to workers – those most endangered by safety offences – in the form of wage cuts or adverse changes in working conditions.

Fifth, fines are wholly counter-productive when they are applied to public sector or government organisations. The effect on such organisations is similar to that set out in the previous point, since, as Fisse (1990) argues, fines in this context simply result in 'some budgetary shuffling with money deducted from one arm of government passing back into general revenue' (cited in Clarkson and Keeting 1994: 243). Since those organisations are funded by public revenues, the effect of fining public sector organisations is ultimately that the costs of organisational offending are transferred to taxpayers.

If the system of fines used to punish corporate offenders for safety crime can hardly be seen to impose a binding penalty on individual firms and public sector organisations, a measure of their more general punitiveness is illustrated neatly by Figure 6, below, which sets out the total value of fines for safety crimes to British industry.

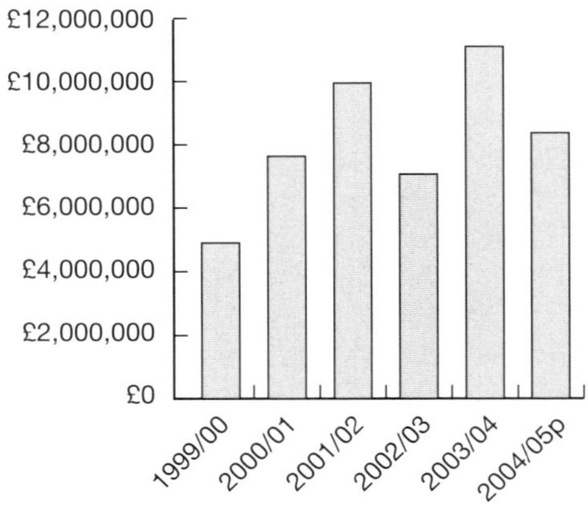

Figure 6 Annual total of all UK health and safety fines[5]

We should read this figure with caution, since the data is from the Field Operations Division of the HSE and although it is the largest division, it represents only around two thirds of the annual convictions in cases laid by HSE and excludes data from chemicals, mining, railways and offshore industries, sectors that have historically attracted higher than average fines. Even with this proviso in mind, though, the HSE FOD data reproduced here is worth commenting upon since the total figure reproduced here does not approach anything close to a sum that might be described as a 'burden' imposed upon business. In fact, the total penalty imposed for safety crimes in the UK in 2004/05, outside of the four sectors noted above (£8,442,340) more or less equates to the combined value of pensions that the top dozen FTSE 100 directors will receive every year when they retire (approximately £8,810,000; Labour Research Department 2006).

Think back to the discussion at the start of this chapter on the key theories of punishment: deterrence, retribution, incapacitation and rehabilitation. Most clearly, the system of fines that we have at present fails as a form of deterrence largely because the penalty is very often, in both absolute and relative terms, set at a derisory level. Relatively low levels of fine mean that this form of penalty fails as an adequate repayment to society for the harm that those crimes incurr. Neither can it be said that fines have the ability to incapacitate offenders, since few are pitched at a level that threaten the continuation of the business. Lastly, the fine as a one-off penalty, unless it has lasting effects (for example if it produces some adverse publicity for the organisation) or is sufficiently harsh, provides no ongoing incentive to organisations to adjust their policies or practices. In other words, fines in this form do not necessarily encourage rehabilitation.

Rethinking punishment for safety crimes

A starting point for thinking about how we might make fines more effective is to return to our five criticisms of the current system of fines outlined above:

1 Unless the organisation faces a high chance of being caught and receives a potentially debilitating fine, there exists no incentive to be law abiding.
2 Fines are not related to the gravity of the offence.
3 Fines do not correspond to ability to pay and therefore tend to penalise the smallest firms.
4 The cost of fines can be absorbed and redistributed by convicted organisations.
5 The effect of fining public sector organisations is counter-productive.

The chapter now turns to an exploration of how reforms to the system of punishment might address each of those criticisms in turn.

Accounting for the chances of being caught

It is important at this point in the discussion to recognise that there is a distinction to be made between theories of *individual* and *general* deterrence. Sentencing can have the aim of either deterring the future conduct of the individual or company being sentenced, or sentences can be aimed at sending a message to companies and directors more generally that criminal conduct will not be tolerated by the state. Given the low rates of detection for safety crimes, the general deterrent effect is crucial for optimising the return on the resources dedicated to the prosecution process and the more general aim of promoting compliance (Bergman and Fooks, forthcoming). In both cases, we might expect the organisation to make a similar calculation that aims to answer the question: is it likely that an offence will be detected at all? In the case of individual deterrence, the question is: will the punishment prevent the organisation from committing the same, or similar, offences again? Either way, the calculation will be made not only on the basis of the severity of the punishment, but also on the chances of being caught. One proposal to deal with problem 1, noted above, then, is to ensure that fines are moderated so as to reflect the chances of being caught. As Etzioni (1993) points out, the detection rate for the crime in question is crucial. As not all crimes are detected and punished, fines should be raised as detection ratios decrease. An organisation might calculate that it has, say, only a one in ten chance of being caught. The organisation may also calculate that the likely fine has a value to the company of the average fine for a safety offence in the courts. But the value of that fine to the organisation, since it only has a one in ten change of being caught, would effectively be a tenth of the value that the courts are likely to impose.[6] Such calculations are of course bound to be based upon rather contrived assumptions and projections. But as we have seen earlier in this chapter, this is very often how organisations, particularly profit-making corporations, do make, or strive to make, decisions about organisational or business strategy. At their heart lies an effort to reach an accurate assessment of the chances of detection, prosecution and conviction. Etzioni presents evidence (ibid.: 153–4) to show that only about one in 50 corporate crimes is detected. Assuming the rational calculation of corporations, fines would thus have to be multiplied by 50.

Fines reflecting the gravity of the offence

Dealing with problem 2 would mean refocusing the way that safety crime offences are generally punished. Specifically, it would mean a shift away from the inchoate mode within which those offences are currently located. As we saw in Chapter 5, an assessment of the gravity of the harm caused does feature in the investigation and in the sentencing process. To argue for a shift towards a consideration of the harm done would therefore merely be to argue that we should move the system a little further in the direction it is already going in. Thus we might recommend a minimum penalty to be linked to an offence in which endangerment was particularly acute or where a minor injury, a major injury or a death had been caused. Those minimum sentences would be applied alongside the existing minimum sentences referred to above. A combination of the gravity of the offence and the chances of being caught might bring the system of fines closer to the kind of rational decision making that might incentivise organisations not to break the law. Rather than moving towards a deterrence model, this approach would be in line with a retributive model, with the penalty fixed to sum that would indicate the appropriate penalty for the offence.

An accurate assessment of the costs of safety crimes to society is, just like any other type of crime, very difficult to achieve. For example, in the case of a street robbery, we may be able to put a price on the loss incurred by the item stolen, but there is also a range of hidden costs: perhaps the victim has been physically or psychologically damaged, for example, and may need medical or welfare support. Safety crimes raise similar problems. The scale and the impact of safety crimes (as we saw in Chapters 2 and 3) suggest that the external costs, those costs never borne by offending companies or their officers, are huge. These include medical costs, funeral expenses, the loss of income to an individual or a family caused by long-term illness or disease, before even thinking about the huge long-term emotional and physical costs that are borne by victims and the bereaved. Once we begin to add those costs on, any fines that could come close to accounting for the true costs of serious safety crimes would most probably be far too high to expect any but the largest of companies to pay.

Unit fines

If fines were reformed along the lines of a retributive or just deserts penalty, then, this would still raise key problems of inequitable sentencing and the likelihood of bankruptcy. It is likely that the smallest companies would be more acutely penalised and the largest firms would be much better placed to absorb and redistribute the costs. An alternative proposal to deal with the problem of uneven impact across large and small companies is to establish a system of unit fines that is linked to ability to pay. This is consistent with current sentencing guidelines. For example, the guidelines issued to magistrates in England and Wales stipulate that: 'In all cases with corporate offenders, the company's financial circumstances must be carefully considered ... Turnover, profitability and liquidity should all be considered ... If the company does not produce its accounts, the court can assume that the company can afford to pay whatever fine it imposes' (Magistrates Association 2001: 3) The problem is that fines are applied highly inconsistently across the country (see Unison/Centre for Corporate Accountability 2002; Hampton 2005). Sentencing guidelines have not in their present form produced consistency.

The CCA has set out a formula for a unit fine that would take account of the gravity of the offence and ability to pay. In this formula, the courts would set a percentage at a level somewhere in the range of 5 per cent to 15 per cent that would reflect the seriousness of the offence. The level would be set according to the degree to which the company was culpable of an offence and the degree of harm caused. Thus, a high degree of culpability and a high degree of harm would place the company towards the higher end of that scale. This percentage could then be applied to an average of either the firm's turnover or profit to determine the level of fine (Centre for Corporate Accountability 1999).

One problem with imposing unit fines is that sentencers currently have very little knowledge about the financial status, or indeed other important aspects of the offender that might affect the sentence. Currently, David Bergman has noted that, when sentencing convicted companies, the courts do not have access to the same level of detail of background information about the offender in comparison to other crimes. Social inquiry reports for the latter are likely to include educational details, income and expenditure. Often inquiry reports will also be furnished with an assessment of the offender's likely response to probation. Yet, in safety crime cases:

> No police officer or similar person gives evidence and there is no document available to the court similar to the social inquiry report. The court remains unaware of the most basic information on the company – its turnover, annual profits, history of relationship with the regulatory agency or its general health and safety record. (Bergman 1992: 1312)

Bergman, in arguing for higher fines, advocates the use of 'corporate inquiry' reports detailing essential financial and safety information. He cites as a model the system in the United States under which a federal probation officer is required to undertake a pre-sentencing investigation into each convicted company to help the court decide an appropriate level of fine.

Equity fines

Problem 4 set out above – the ease with which the cost of fines can be absorbed and redistributed by convicted organisations – raises the question that setting an appropriate level of fine for a company may be futile in any case, since managers and directors can simply pass on the burden to consumers, shareholders or workers, especially in the largest firms. One solution developed by Coffee (1981) is to impose fines upon the value of the firm, rather than upon its running costs. This proposal for 'equity fines' proposes that offending companies would be ordered to issue a set number of new shares in the firm, to be controlled by a state-controlled compensation fund. The argument is that this process, which effectively dilutes the value of shares held by the owners of the company, would not penalise the most vulnerable groups, since the funds for investment would not be depleted, merely re-allocated to the compensation fund from existing shareholders.

The major criticism of equity fines is that those who are penalised, the shareholders, are likely to have little knowledge of the offending and are not in a position to influence the day-to-day decisions made by a company; equity fines therefore target innocent shareholders (see, for example, Croall and Ross 2002: 541). This criticism is countered by the argument that it is ultimately shareholders who benefit from the proceeds of crimes committed by the company in their name. Further, a system of equity fines should encourage shareholders to take an active interest in the degree to which their firm complies with the law and pressurise the company to ensure it is law abiding. To this end, Glasbeek (1989) has suggested that equity fines might be strengthened to apply to major shareholders: those shareholders with sufficient equity to exert substantial influence or control over the affairs of the company. A formula to define who might be considered a major shareholder, Glasbeek notes, is already applicable in existing corporate statutes in Canada. Such a formula, he suggests, could also be applied to parent companies and holding companies where it is a subsidiary that has committed the crime.

To this, we might add, there is no reason why equity fines could not be based upon the type of graduated formula that we describe above; thus they could be set at levels that reflect the seriousness of the offence and the ability of the company to pay.

Beyond financial penalties and 'pure' deterrence

The fifth problem, that of sentencing organisations that are in the public sector, immediately demands that we need to look beyond the realm of financial penalties, simply because of the contradictions that are raised by state judiciaries punishing public authorities when the costs will be paid for out of public funds. In any case, if our aim is the reform of organisation – and the provision of incentives to organisations to be law abiding – then perhaps we need to be thinking more imaginatively beyond the realm of deterrence.

Deterrence has always been attractive to white collar and corporate crime scholars, simply because the positivist assumption that lies behind deterrence theory (that individual action is shaped by a rational calculus and therefore produces a relatively easy solution to the crime problem for which crime experts can take credit) proposes similar remedies for crimes in the suites as mainstream criminology has proposed for crimes in the streets: all we have to do is remove the gain to be made from crime and we will eradicate the crime. Unfortunately there are several reasons to expect that things might not quite be as easy as this.

First, decision makers work within what has been called 'bounded' rationality. This means that other factors beyond economic calculus tend to shape their decisions. They may be influenced, for example, by what their peers expect of them or by what they perceive as normal practice in an industry, or they may make decisions for very specific reasons that are related to the structure or practices of the organisational bureaucracy (Bergman and Fooks forthcoming). This suggests that financial considerations might never fully capture the decision-making process.

Second, even if we could assume that decision makers even in the most profit-orientated organisations do structure their actions and inactions according to a carefully worked through calculus, then because of the problem of identifying the true costs of crime captured by externalities, this calculus could never be expected to produce a 'pure' form of deterrence.

Third, and following the previous point, we should be wary of relying upon financial costs, particularly since the actuarial science of cost-benefit analysis can never be an exact one. The rational calculations made by corporate accountants do not always produce rational results (as we saw in the examples of Ford and Railtrack cited above). The risk is that an organisation's or its senior management's commitment to the law is reduced to a figure on a balance sheet by financial penalties, regardless of whether this figure is accurate or not (on this point, see Becker 1968).

Fourth, a problem arises that Coffee (1981) has described as 'the deterrence trap'. Introducing fines that would be large enough to truly act as a deterrent would mean that most corporate offenders, if they were caught, would simply end up bankrupt. For Coffee, this is pointless since those that are penalised by bankruptcy are generally not those that are responsible for safety crimes. It is workers through job losses and clients and contractors through unpaid bills that are generally penalised by bankruptcy. This effect, termed 'spillover' by Coffee, is one of the reasons that led him to develop his proposals for equity fines, but it also gives us reason to think that non-financial penalties might be necessary when punishment is aimed at reforming rather than destroying corporate offenders. Having said this, there may well be firms or organisations that society deems so beyond reform and so criminogenic that we would not wish them to continue in existence. In such cases, several solutions have been proposed.

Incapacitation

The corporate death penalty is the ultimate sanction against an organisation. In cases where the organisation is put to death, companies would effectively be nationalised or put into the hands of a receiver (as is the case in bankruptcy procedures in many jurisdictions). Braithwaite and Geis (1982) argue for sentences which might fall short of the corporate death penalty, such as limiting the charter of a company to prevent it from continuing the criminal aspects of its operations, or withdrawing its licence to operate in a particular sphere of activity. The option to nationalise is attractive because it can limit the spillover effect of job losses or economic harm to a particular section of the community. In other words, it avoids victimising the most vulnerable populations. The transferral of a company to a new parent company with an established law abiding record might provide an option here. A corporation may also be barred from engaging in a particular form of economic activity or prohibited from operating in particular geographical regions (Moore 1987: 395–6). A similar and less permanent option that has been suggested has been the freezing or confiscation of the assets of a company.

Similarly, it is common in other types of white collar crimes that senior managers and directors are disqualified from holding high office, either permanently or for a set period of time. This would seem to provide a means of incapacitating individuals from committing similar crimes in the future. The weakness of using this resort exclusively is that the organisation can easily find someone to take the place of the disqualified director; the removal of one or more individuals will not necessarily reform organisational behaviour. As Levi (2000) has pointed out, it is also the case that this can be relatively easily circumnavigated by individuals who have a controlling position in their own firm, since they may simply establish a partnership (rather than an incorporated firm) and continue to conduct business.

One final means of incapacitation is to exclude companies – and directors – from bidding for central or local Government contracts. The role of government as a customer of private business firms is significant and, following the recent waves of privatisation in the UK and beyond, a growing one; thus governments have considerable leverage that could be wielded in order to exclude those with convictions, or poor safety records, from bidding for, or receiving, Government work; or, somewhat less bluntly, such information could at least be used as one factor to discriminate between bids. This would both operate as a form of disqualification and potentially force companies to take a future-oriented, bottom line view in the day-to-day management of safety and compliance with legal requirements.

Rehabilitation

Lofquist (1993) draws an important distinction between 'market-based sanctions' and 'politics-based sanctions'. The former are founded upon the assumption that corporations are unitary, rational profit-maximising entities, and that crime results from these characteristics. So, *corporate* crime control might be more appropriately achieved by increasing the costs of safety crimes above their potential benefits. The latter are founded upon the assumption that organisations are complex, differentiated entities, simultaneously in pursuit of different and often conflicting goals. Safety crimes can therefore be understood as emanating from the structural characteristics of the organisation. Punishments that are aimed at altering organisational structures and procedures in order to improve internal accountability and law observance might therefore be more appropriate (ibid.: 165; see also Piquero *et al.* 2005). Some have argued that punishments are more feasible and more likely to be effective if their aim is the reform of the criminogenic structures, policies and practices of profit-making organisations. As we have seen in Chapter 6, the focus upon the criminal features of the organisation is an approach that has been adopted in Canada and advocated in Australia, where in the case of the latter, the national penal code allows for prosecutions on the basis of a criminogenic 'corporate culture'. We might in some cases expect responses that seek to change the corporate culture, or organisational rehabilitation, to be more valuable than a simple deterrent effect.

The concept of organisational rehabilitation, however, raises a series of other issues. As the case studies in Chapter 1 demonstrate, safety crimes very often arise from defective control systems, insufficient checks and balances within the organisation, and poor communication systems or authoritarian managerial regimes. These failings are sometimes deliberate, and made to facilitate the commission of offences or the avoidance of detection, and sometimes the failings are inadvertent. Either way, it may be possible for court orders to force corporations to correct criminogenic polices and practices. Rehabilitation, according to Braithwaite and Geis, is a more workable strategy with corporate crime than with mainstream crime because organisational structures are more easily reformed – they are more malleable – than human personalities. As they argue, 'a new internal compliance group can be put in place much more readily than a new superego' (1982: 310). Moreover, Croall (2005) points out that since corporate penalties are a response to corporate 'fault', it is appropriate that penalties are directed at the underlying circumstances of this fault in the organisation. Corporate probation, she suggests, is likely to be effective in this respect.

Corporate probation, an established punishment used in the US courts, involves imposing a specified set of conditions on the convicted company. Corporate probation might involve, for example, the introduction of specified safety procedures and/or the employment of specialist safety staff. The first time probation was used against a corporation in the US was in 1971 where the oil company ARCO was ordered to develop an oil-spill response programme during its probation period (Lofquist 1993). Under US sentencing guidelines, certain mandatory conditions of the probation are now specified. If a company commits a further crime during probation, it must pay a fine or restitution or it must engage in a programme of community service. In the US system, regulatory officials act as probation officers and report to the court on whether or not conditions of probation have been satisfied. Court orders, known as 'punitive injunctions', have been proposed as a more stringent form of corporate probation. The point of punitive injunctions is that the form of remedial action required by the courts is linked very closely to the process of triggering a criminal penalty if the injunction is breached. Punitive injunctions tend also to specify the personnel responsible for ensuring their compliance (Croall 2005).

In the US, offenders' rehabilitation programmes for senior officials in the organisation who are convicted of safety crimes normally involve a range of professionals who encourage the offender to face up to the impact and thus understand the consequences of his/her behaviour. However, the use of such programmes can be questioned since, if we accept that often the senior officers who commit safety crimes are likely to be in a much better position to appreciate the consequences of their actions for themselves and for others (they would for example normally be sufficiently expert to appreciate, prior to the offence, exactly what damage, loss or injury could follow from their wrongful conduct and why it is wrong), then such programmes may be of limited value.

Restitution for safety crimes

Some commentators have noted the possibility of community service as a means of corporate recompense for the damage caused to the individual, the family and the community. Punch (1996) has pointed out the advantages of corporate community service (for example making a company lend an executive to a charity for a year), and direct compensatory orders which enable the punishment to be designed to offset the costs of the crime to the victim. Corporate community service is a remedy that is widely used in the US courts, as is the imposition of community service upon individuals associated with organisational crimes. Often the courts design community service in a way that allows the organisation or individuals within the organisation to contribute to the community using their specialist skills. It is this latter feature of community service for safety crimes that has often exposed such penalties to criticism for their inappropriateness. Thus, for example, in August 2006, a former President of a sewer company was sentenced to seven years probation and 840 hours community service by an Arizona court after he had been found guilty of homicide and reckless endangerment for causing the death of an employee, James Gamble, who was poisoned by sewer gas whilst unblocking a sewer tank. The court found that the main cause of the death was that the air had not been adequately tested, and crucially that workers were not properly trained in the correct procedures for entering sewer tanks or in the proper rescue procedures. The community service, the court ordered, should be spent teaching safety classes. This caused controversy since the victim had died precisely because the company had failed to provide the necessary safety training. Critics argued that the company was clearly not capable of providing this training to the necessary standard.[7]

Shaming provisions

Public shaming provisions, such as publishing an advertisement in a newspaper that publicly announces a conviction for safety crimes, or being ordered to show a sign detailing the conviction outside a firm's premises, have been used in several jurisdictions.

For Braithwaite and Geis (1982: 301), those punishments can act as a deterrent simply because '[c]orporations and their officers are genuinely afraid of bad publicity arising from their illegitimate activities'. Publicity orders by their nature are dependant upon how their message is communicated and the medium through which this message is communicated. This might be particularly important when they are applied to safety crimes, since publicity orders depend upon public reaction. Public reaction against some safety crimes – those that involve no injury or death particularly – may be more difficult to secure precisely because of processes of decriminalisation that safety crimes are vulnerable to (Dunford and Ridley 1996: 15). Communicating a strong message of the seriousness of the offence may be all the more important in the context of safety crimes. As Levi (2002) has noted, this type of penalty is likely to be more effective when it is applied to large household name companies, since they are the companies that have the most to lose in terms of reputation. However, it is significant in this respect that Fisse (1971) argues that publicity orders and similar techniques of shaming rely less on public reaction than on the reactions of business executives, official persons and 'opinion leaders'.

Publicity orders can impose punitive burdens on management which are more difficult to quantify and therefore less calculable than fines. Ultimately, publicity orders are more likely to affect the organisation's reputation than have a lasting impact on a chief executive or a director. Therefore, those are punishments for which the costs will always be disproportionately borne by the organisation rather than by individual senior managers. A related requirement that has been used in the US in particular is the requirement that a senior manager/director appears in court during the trial and sentencing process where it is an organisation that is being prosecuted so that the company or organisation's senior officers are not allowed to use the shell of the organisational form to remain anonymous.

The requirement for a human face to be given to the corporation in court is significant since it takes a major leap of the imagination to conceptualise how the corporation can exhibit an emotional response such as shame. The internalisation of shame by the corporation rather than the individual is more likely to be realised in a completely different way and it more likely to be internalised as a measure of the extent to which bad publicity costs companies in the long term. Shaming itself may therefore be regarded as a financial deterrent to companies as opposed to an emotional response capable of allowing individuals to contemplate the consequences of their offending. It may also be the case that precisely because they are less quantifiable than fines and the fact that their long-term effects are less amenable to quantification, reputational penalties can impose greater punishment and deterrence than mere economic sanctions.

Friedrichs (1998: 13) however, has argued that 'sophisticated public relations operations have the means to transform attempts to shame corporations to their advantage, or at least to neutralise those efforts'. In other words, this is a form of punishment that may be easily countered by the larger corporations with large public relations departments, or those that can afford to hire PR firms. At least two further problems are associated with this sanctioning strategy. One is that it requires firms to have some visibility – or indeed engage in direct business with – the general public. Yet often the public fails to identify which firms are the source of goods or services they might purchase – how many of us would be aware, for example, that Dixons is actually part of a European multinational company that includes Currys, PC World and the Link, companies to whom we might as consumers direct our business were there to be any reaction to the public 'shaming' of Dixons. Second, even in the event of effective shaming, if the impact of reputational damage is to lose custom, then this requires choice – however, for those who buy petrol, and might want to respond to the public shaming of a leading retailer, they would find not only little choice (petrol all emanating from one of the big five oil giants), but they would also find there is little to choose between any of these in their long-term safety management/compliance record, for example.

Towards a sentencing mix?

Differentiating between those different aims of punishment (deterrence, retribution, rehabilitation and incapacitation) does not imply that those philosophies of punishment produce mutually exclusive solutions to the punishment of safety crimes. As the preceding discussion indicates, there are likely to be cases where deterrence is not likely to be achieved (not least if punishment comes be regarded simply as the 'add-on' cost of conducting business); equally, there are likely to be cases where the most appropriate 'just deserts' is impossible to establish, or where rehabilitative aims are simply not appropriate. If we imagine that the penalties imposed for safety crimes might not be merely restricted to a financial deterrent, then there is reason to think that we can combine deterrence and just deserts to good effect. But because the real value of the external costs generated by safety crimes is unlikely to be quantifiable, we will always need to have the option to either seek rehabilitation or, in the worst cases, incapacitation. This suggests that the most effective way of dealing with safety offences is by deploying a 'mix' of sentencing options that could be made available to the courts (Croall and Ross 2002). In other words, because of the complexities of the way that safety crimes are produced and the complexities that we face in seeking to prevent future crimes taking place, we have to think about the most appropriate 'sentencing mix'. This is not to say that each punitive aim should be given equal weight. Indeed,

Braithwaite (2002) has suggested an adaption of his pyramid of enforcement along the lines of a sentencing mix. Here, he recommends restorative justice as the first resort at the bottom of the pyramid, then if this fails, regulators should appeal to the rational capacities of offenders and impose deterrent penalties. Where those fail and the offender is deemed an 'incompetent or irrational actor', then incapacitative penalties should be imposed.

All of the evidence that we have points to the relatively high chance of securing deterrence where financial penalties are appropriately apportioned and where those penalties are most clearly targeted (for detailed overviews of this evidence, see Davis 2004 and Bergman and Fooks forthcoming). For this reason, we would argue that a system of fines that reflects both the gravity of the offence and the ability of the offender to pay seems to be the most justifiable and the most likely to achieve deterrence. In cases where the organisation has the necessary capital to sustain equity fines, we see no good reason why this penalty should not be levied on investors. The fact that the punishment of investors is hardly likely to find favour in capitalist social orders, where a range of legal protections are already afforded to the owners of businesses (and here we refer the reader back to our discussion of corporate personality in Chapter 6) does not mean that we should not continue to argue for equity fines as a likely means of achieving results. After all, we do not consider the interests of those that profit most from law breaking when framing punishments for any other crime.

It is also important not to underestimate the importance of attaching moral condemnation to punishment and the way that this might supplement purely instrumental punishments such as effective fines or reforming the structures and policies of the organisation. As Bergman and Fooks have noted, the key weaknesses of those instrumental approaches is that they fail to communicate and transmit the requisite moral opprobrium that is central to the process of criminalisation. Publicity orders thus are important as supplementary penalties, since they provide a different function: 'as a sanction that can work to reaffirm the moral importance of meeting the terms of legal duties under health and safety legislation, publicity orders represent a potentially powerful means of reinforcing the normative basis of compliance with health and safety regulation' (Bergman and Fooks, forthcoming: npn).

Sentencing individuals

Thinking about the most appropriate 'sentencing mix' also returns us to the question of whether we punish the individual or the organisation. If this chapter has concerned itself almost exclusively with corporate and organisational penalties, this is merely because of the infrequency with which individuals are punished for safety crimes. Yet, again we might raise issues about the extent to which this is driven by the discretionary decision of investigators and prosecutors. HSE prosecution policy currently notes that the regulator will seek to prosecute individuals using the same evidence test noted above, and instructs inspectors that '... prosecuting individuals will be warranted where there are substantial failings by them, such as where they have shown wilful or reckless disregard for health and safety requirements, or where there has been a deliberate act or omission that seriously imperilled their health and safety or the health and safety of others' (Health and Safety Executive 2003c: 2). The policy also emphasises that decision to prosecute also rests upon a 'public interest test', so that investigators must ask themselves: would prosecution be in the public interest? Either due to the evidential test, or the public interest test, or a combination of the two, however, individual prosecutions are relatively rare. For the individual the maximum penalty for a summary health and safety offence in the UK is six months imprisonment or a £5,000 fine or both. Yet, as we noted in Chapter 6, it remains overwhelmingly the company that is punished for safety crimes. To recap, we reported that in the ten years up to December 2005 only 86 directors were convicted and 11 jailed for breaches of health and safety law. We also reported that between 1980 and 2004, 11 directors were convicted for the common law offence of manslaughter. Of those, five were imprisoned, one received a community sentence and five were given suspended sentences.

Fisse and Braithwaite (1993) argue in favour of corporate criminal liability but are also conscious of the need in many cases to punish or discipline individuals within the corporation. Geis and Dimento (1995: 84) conclude that punishing the individual *and* the corporation offers the best chances of securing deterrence, but speculate that results 'might more readily be achieved by concentrating all criminal resources on individual malefactors'.

It is at the punishment stage of the criminal process that the *de facto* corporate veil discussed in Chapter 5 comes to life. For if it is companies rather than their senior officers that pay the penalty for committing crime, the only penalty to senior officers is that their reputation as managers may be damaged. But if the individual responsible is not prosecuted for, or directly implicated in, the crime, then this is hardly likely to affect their managerial reputation. It is equally arguable that imprisoning directors will not really make an impact upon levels of safety crime because companies will always be able to find someone willing, in exchange for the right sort of remuneration, to take very high-risk decisions. The US phenomenon of appointing a 'vice president in charge of going to jail' (Braithwaite 1984) thus threatens to undermine the effectiveness of punishment. If we accept the principle of *corporate* or *organisational* punishment, a principle that allows organisational offences to be regarded as conceptually independent from the offences committed by the individuals within the corporation, this does not mean that we are left with an either/or choice. Segregating organisational from individual punishments merely gives us the opportunity to use either or to use both. As we have seen, there is reason to think that neither on their own can be effective and, following a long-standing tradition of research in white collar and corporate crime we would argue in most cases an appropriate sentencing mix directed against both corporations and individuals will be the only way to make the punishment stick.

Conclusion

As the previous section in this chapter has argued, the complexities of the motivations and factors that encourage safety crimes demand an appropriate punitive 'mix'. This is not to undervalue the centrality of punishments that aim to provide a necessary deterrent. Safety crimes arise very often from reasoned and cost-balanced decisions. The principle of deterrence therefore must be placed at the heart of any punitive strategy if it is likely to have any effect in preventing future offending at all. Punishment in other words, where rational decision-making has resulted in a crime, has to be constituted in a way that makes crime unattractive to offenders.

It is the principle of 'less eligibility', or the idea that the material conditions of an offender's punishment are regulated to a level far below the conditions experienced by the average non-criminalised citizen, that has historically underpinned the application of deterrence theory (Rusche 1933/1980). Since the imprisoned population tends to be overwhelmingly comprised of those from the lower social stratum, the conditions of imprisonment must be seen to impose more hardship than the social conditions normally experienced by the lowest strata of society. This principle of less eligibility, as we have seen, does not apply to offenders associated with safety crimes (also Munro 2006). Ruche and Kirchheimer noted in their classic work *Punishment and Social Structure* that the general principle of ensuring that punishment negates the proceeds of crime was never applied in relation to the regulation of factory safety (1939/2003: 174), and they reproduce data from French labour law violations to demonstrate this point. Their argument is that the system of punishment encourages rather than discourages offending in the case of safety crimes. As we have seen in relation to our discussion of the form that punishment takes in the case of the UK, their basic argument remains prescient. As Rusche (1933/1980) has argued, the existence of a principle of less eligibility is necessary because the form that punishment takes is always related to structural inequality.

Punishment is less related to crime, as we have already seen in our discussion of the historic origins of the criminal law, than to the defence of a particular social order, one that is based upon unequal relations of property (see also Rusche and Kirchheimer 1939/2003: 5). Safety crimes, just like many forms of mainstream crime, ultimately arise from unequal social relations of power. As we have seen in Chapter 2, those are offences that tend to victimise disproportionately the most vulnerable groups in society precisely because of their subordinate position in the labour market, and more generally, the social hierarchy. In Chapter 6, we argued that the processes of law reform and implementation of the law are intimately connected to a process of social ordering; a process that is geared towards maintaining the status quo. In capitalist social orders, maintaining the status quo means ensuring the predominance of capital and therefore the right of its institutional form, the corporation, to remain profitable. To this we can also add that maintaining the status quo ultimately means ensuring that other key institutions such as branches of the state and public sector enterprises are not fundamentally undermined by the criminal law. It is those rules of thumb that have historically ensured that safety crimes – and the forms of sanction that are used to punish them – are differentiated from 'real' crimes. Just as the magistrates of the nineteenth century found ways of reconstructing offences against the Factory Acts as less blameworthy, a range of judicial techniques continue to undermine the seriousness with which the courts deal with safety crimes (Bergman and Fooks forthcoming).

So what does the preceding discussion tell us about the form that punishments for safety crimes should take? First that the punishment of safety crimes, if it is to be effective, must overcome the major contradiction that we point to above: it must challenge, rather than reinforce, the conditions that give rise to safety crimes in the first place. Some of the suggestions outlined above seek to do so, but ultimately criminal law and a system of punishment cannot achieve the broader aim of social (as opposed to criminal) justice. The key problem with relying merely on punishment as a panacea for crime is that the conditions that give rise to crime in the first place cannot be changed by the way that we punish the small minority of offenders that end up in court. Punishment can never, therefore, be seen as the solution to crime.

Some forms of punishment can provide a general deterrence against particular forms of offending if organisations and their senior officers calculate that it will not be worth their while to continue to commit safety crimes. Punishment in the courts can also send an important expressive message from the state to assert social disapproval of such offending.

Punishment can also be targeted so as to produce less tangible, but nonetheless effective results, such as reinforcement of the moral importance that should be attached to safety crimes. There are also, as we have seen, punishments that can go some way to repairing the social damage that a particular crime has inflicted. Finally, as we have argued here, the particular forms that punishment take are likely to be a great deal more effective if they are designed to take into account the specificities of the crime and the offender in safety crimes (not least a recognition that they are very often highly profitable crimes conducted from behind the corporate veil). It should also be added that the discussions that we have had in this chapter are not merely theoretical, since versions of each of the punishments discussed in this chapter (with the exception of equity fines) have been, or are currently, in use in the US, Canada, Australia or Europe (ibid.).

Indeed, any review of both literature and international practices in relation to the sentencing of corporations and/or individuals involved in corporate crime more generally attests clearly to the fact that this is an area in which there now exists a range of imaginative proposals, some of which have been introduced in limited fashion, others of which remain at the proposal stage (Croall 2005). These sanctions each have their drawbacks, none is a panacea, and each is more or less appropriate for particular types of corporation and following specific forms of offence (Slapper and Tombs 1999). Ultimately, while there are clearly enormous difficulties in developing effective sanctions in the case of corporate crime these tend to be political rather than technical (Lofquist 1993, Etzioni 1993). Again, the key issue is the political refusal to treat corporate crime in general, and safety crime in particular, as real crime.

This reference to the politics of safety crime reminds us that, while the importance of the forms of punishment adopted by any given society should not be underestimated, the key features of societies that produce safety crimes remain, as this book has indicated, located in the organisation of production (Chapters 1, 2 and 3) and the political construction of the form of regulation designed to control safety crimes (Chapter 7), rather than the outcome of the criminal justice process. It is those central features of the production and control of safety crimes that urge us to reconsider the centrality of political economy, rather than the peripheral amelioration of particular crimes that might be provided by reforming systems of punishment. It is to a development of this political economy approach to understanding the production of safety crimes that our concluding chapter turns.

Notes

1 A practice for which there are clearly limits; see, for example, Tombs 2002.
2 Source of data is Health and Safety Executive (2005).
3 Source of data is Health and Safety Executive (2005).
4 *R v Howe and Son* (Engineers) Ltd [199] All ER 249.
5 Source of data is Health and Safety Executive (2005).
6 It is worth noting in the context of our discussion above that if the chance of being prosecuted for a detected death is 29 per cent, and the number of detections is between $\frac{1}{7}$ and $\frac{1}{8}$ of the true figure of people who are killed by working, then in fact the chances of being prosecuted for killing a worker, never mind for less serious offences, is likely to be a great deal less than one in ten.
7 See www.occupationalhazards.com/articles/15516.

Chapter 9

Conclusion: making sense of safety crimes

Introduction

Safety crimes are a significant crime problem, as we have sought to demonstrate throughout this text. Yet, despite this, we have also indicated various ways in which political, social, legal and regulatory processes combine to obscure their nature, extent, scale and consequences. This social construction of safety crimes as something other than a crime problem extends to the discipline of criminology which, through its definitions of 'crime', 'violence' and 'policing' further marginalise such phenomena – yet a central theme of this book is that safety crimes are a proper and legitimate focus of analysis for criminology. Indeed, it is a remarkable feature of criminology that this is the first text in the discipline devoted to this subject. It remains for us to consider the extent to which, theoretically and empirically, safety crimes might be placed upon the criminological agenda.

We saw in the previous chapter how there are good theoretical and empirical reasons for at least considering the application of classicist or 'rational choice' forms of reasoning to the area of corporate crime control. Whilst the rational choice perspective pre-dates modern criminology, as we have noted, it has made an indelible mark upon the development of deterrence perspectives. To argue, as we have, that rational choice theory is more applicable to safety crimes than to the types of crimes to which it is normally applied, assumes broadly two things about safety crimes. First, that safety crimes result from profit-seeking (see, for example, Grayson and Goddard 1976; Nichols and Armstrong 1973; Woolfson *et al.* 1996; Work Hazards Group 1988). And, second, that safety crimes are produced through the functioning of rational, profit-maximising entities.

In for-profit organisations, the claim that accumulation (to sustain or expand profitability) ultimately takes priority over safety – indeed, any other goal – within a corporation seems incontrovertible. Thus it would be ludicrous to ignore the dynamic tendency to accumulate within a capitalist system, since this provides the *raison d'être* of the private corporation. However, the primacy of accumulation does not mean that there cannot be some congruence between occupational safety and efficiency/profitability within a given organisation. There is no need to accept uncritically a business case for improved safety[1] or to accept the tenets of consensus theory (see Chapter 7) to recognise that, under certain conditions, there is some coincidence of interests between accumulation and safety, and between capital and labour. To pose a mutual exclusivity between safety and profitability leads to a misunderstanding of the causes of industrial injuries in general, and safety crimes in particular, and – crucially – an under-estimation of any prospects for their prevention. Such a view for example fails to account for the fact that some companies manage safety much more effectively than others, and are also in some cases among the world's most profitable corporations, and that safety measures over and above those required by the law are taken by some companies at certain times, and so on. Moreover, the safety–profits dichotomy downplays important points about the way in which profitability is calculated. For example, a corporation that engages in long-term calculations of profitability is more likely to provide safer workplaces than one driven by short-term rationalities. Equally, large organisations engaged in a range of more and less hazardous activities are able to subsidise improved safety in some aspects of their operations, and even to sustain loss-making units, for some (albeit not indefinite) period of time (Pearce and Tombs 1997). Finally, some versions of the safety–profits dichotomy tend to pose an overly rationalistic concept of corporate managements, safety efforts, and the production of industrial injury. In other words, there is an implication, at least, that deaths and injuries can simply be attributed to acts or omissions based upon (financial) calculation. While this may be so in some cases, it is clearly not applicable to all safety crimes. Moreover, to attribute rationality to the corporation is to recognise that this is its *raison d'être*, rather than a description of how real companies actually operate at all times; as we have seen in our case studies, managements often manage poorly, so that calculations are either not made or, if they are made, are in fact erroneous. It should also be added, of course, that safety crimes may also be produced in the context of not-for-profit organisations (see Chapter 6).

For these reasons, then, while there is an ultimate and inevitable 'truth' to the argument that profit maximisation within capitalist economies is the most fundamental cause of safety crimes, we need to move beyond this level of analysis if we are to explore fully how safety crimes are produced. It is to the use – the potential and limitations – of criminological theory (beyond the rational choice perspective) in furthering this understanding, that we now turn.

Criminology, theory and safety crimes

Various forms of individual positivism that emerged after the heyday of the eighteenth- and nineteenth-century classicist theorists sought to identify the 'abnormalities' that either propelled individuals into crime, or ensured that they were more predisposed to committing crime than the general population. Now, there have been some attempts to apply an individual positivist type analysis to corporate crime, though these have been marginal to criminology, conducted mostly within business or management studies. These have tended to seek to identify those 'personality' factors associated with people who succeed in private companies, and tend to highlight features such as being innovative, ambitious, shrewd, aggressive, impatient, and possessing a 'moral flexibility' (see Snider 1993). More recently, however, and squarely within a 'criminal justice' framework, Babiak and Hare (2006) have examined the role of the 'psychopath' in corporations. As Hare has claimed,

> The world of unfeeling psychopaths is not limited to the popular images of monsters who steal people's children or kill without remorse. After all, if you are bright, you have been brought up with good social skills, and you don't want to end up in prison, so you probably won't turn to a life of violence. Rather, you'll recognise that you can use your psychopathic tendencies more legitimately by getting into positions of power and control. What better place than a corporation? (Hare, cited in Hilpern 2004)

Now, we do not need to accept pseudo-scientific categories such as 'psychopath' (see also Bakan 2004), nor the theoretical framework of individual positivism, to see how certain qualities are likely to be both valued within the corporate world, while at the same time individuals possessing such characteristics may also be more likely to be involved in corporate illegalities, either as leading figures or as individuals prepared to turn a blind eye to organisational illegality.[2] If corporations seek to recruit particular types of people, and if it is the case that the higher one goes up the corporate hierarchy the more likely are certain characteristics to be present, valued and accentuated, then we need to know something about the culture and functioning of the corporation itself, as well as the environments within which it operates, to understand how its employees, from the most senior downwards, act, think, rationalise, and so on. For example, to understand how Bhopal was produced requires more than understanding the actions and omissions of Warren Anderson; on the other hand, Anderson clearly typified many of the qualities valorised by corporate America, and thus requires some place in an overall story of the production of thousands of deaths. In this context, it is important to bear in mind that very often there is an almost total lack of meaningful differences between corporate offenders and corporate non-offenders (Snider 1993: 61; see also Virta 1999; Weisburd and Waring 2001).

If a myopia towards corporate offenders has been a hallmark of individual positivisms, so too has this been the case with the vast range of sociological positivisms that have come to dominate criminological theorising since the 1930s. Here, however, we come to some notable exceptions. As a criminological theorist working within the Chicago School, Sutherland himself attempted to develop a general, sociological, theory of crime causation, claiming that 'differential association' could explain both upper-class and lower-class crimes: crime arises from an excess of definitions favourable to law violation over definitions unfavourable to law violation. Criminal activity – motivations, *post-hoc* rationalisations and actual techniques of commission – is, like all behaviour, learnt. This learning, and exposure to different definitions regarding the appropriateness or otherwise of certain behaviours, emerges out of our various associations – and these associations vary by frequency, duration, priority and intensity.

Moreover, we know, on the basis of documented evidence, insider accounts, and, indeed, reasonable inference, that within certain corporations or even industries, certain forms of activity are prevalent, both in terms of knowing how to engage in them and knowing why one must engage in them. So presumably if this holds for legal activity, it holds for illegal activity too. For example, in the cockling industry off the north-western coast of England, one must assume that 'everyone knows' that there is available a pool of illegal labour, and how to draw upon this, just as in the construction industries of many of our regenerating urban centres, it is common knowledge that there are armies of migrant labour, where the pick up points and times to collect these each day are, and how these should be treated to conceal them from any external authority, and so on. Further, and crucially, there may also be generalised knowledge within a particular sector that 'everyone is doing it' – which not only provides a motivation, since not to do it is to place one's own workplace or company at a competitive disadvantage, but also that to do it is so generalised that it is acceptable, not really criminal.[3]

Differential association is a problematic concept, and has been subjected to stringent criticism (Taylor *et al.* 1973: 125–30). It is of interest, however, precisely because it attempted to incorporate corporate crime within a general theory of crime. Other variants of sociological positivism have not sought to do so in such an explicit manner, yet there still remain elements or forms of these modes of explanation that can be or have been utilised by subsequent theorists to explain incidences of corporate crime. Notable here is Mertonian strain theory and its central concept, anomie. For example, Passas (1990) has linked deviant behaviour to the disjunction between institutionalised aspirations and the accessibility to legitimate opportunity structures. Although Merton saw these phenomena as confined to the lower classes, Passas argues that there is no compelling reason why anomie theory cannot be applied to high-class and corporate deviance:

> As the meaning and content of success goals vary from one part of the social structure to another, similar difficulties in attaining diversely defined goals may be faced by people in the upper social reaches too; they are, therefore, far from immune to pressures towards deviance (ibid.: 158).

Of course, the pressures to succeed exist for businesses and organisations in terms of maximisation of profit, growth and efficiency. These goals may have to be obtained *by all* or *any means*, particularly when the continuation of the corporation is at stake and key actors have come to equate the furthering of their own ends as largely dependant on the prosperity of the firm, an attitude underpinned by the system of financial rewards which apply to senior management, not least bonus and share schemes linked to stock market performance. Structural pressures and strains may be applied both to those at the top as well as to employees, and the employment of deviant methods may be the only possible way of dealing with problematic situations, or may be perceived as such (see Box 1983).

Finally, in this whistle-stop tour of criminological perspectives, various forms of critical and radical criminologies – including Marxisms and feminisms – have made important contributions to our body of knowledge regarding corporate crime causation. Notable here are: Stuart Hills's edited collection *Corporate Violence* (Hills 1987b), precisely because it consists of a series of empirical and theoretical case studies of the ways in which injury and death are produced systematically by the drive for profit; *Crimes of the Powerful* (Pearce 1976), where it is argued that corporations act systematically to control the markets within which they operate, often criminally, and often doing so with the connivance of the (here, American) state; and Szockyi and Fox's (1996) anthology of analyses of the myriad ways in which corporations exploit constructions of gender to victimise female consumers, workers and recipients of health care.

The processes which result in deviance and anomie thrive in an environment dominated by concerns of costs and benefits and are reinforced by the structural pressures generated within a capitalist mode of production. Sometimes the connections between market pressures and corporate crime are easily seen. Companies often end up committing crimes or making terrible reckless errors because they are desperately endeavouring to acquire, retain or re-gain a significant slice of the market, as was clearly the case with the *Herald of Free Enterprise.* Thus, as we have seen throughout this text, but especially in relation to the case studies in Chapter 1, there are clear structural pressures upon and within organisations which seem to provide a crucial motivation for safety crime – as a desirable, or even 'necessary', response to the demands of profitability. This opportunity-motivation is clearly heightened when these structural pressures are faced in contexts where workers are relatively weak or vulnerable.

Beyond criminology

One obviousness that emerges from this brief appraisal of criminological theory, and indeed throughout the text, is that understanding safety crimes requires us to look far beyond the individual on the scene – the ship's assistant bosun who did not check whether the ferry's bow doors were closed before leaving port, the chemical plant worker who attached the hose to the tank, the shipping manager who sends a young man to unload cargo. This is to deny neither human agency, nor that individuals at times act (or fail to act) in ways that are decisive triggers for safety crimes. Individual, low-level employees at times may act in violation of safety law, behave stupidly, carelessly, recklessly, and so on. Having said this, as we have argued throughout this text, and in particular in Chapter 3, the extent to which safety incidents and safety crimes are represented as being the effects of one or several low-level individuals acting in such ways is both enormously exaggerated, and indeed ideologically and strategically so.

To examine incidents and offences in terms of individuals is to fail to ask the question, what kind of organisation or process is it that allows the actions or inactions of one or several low level employees unintentionally to cause significant physical harm, as is often the case? Here, we can usefully turn to Charles Perrow's (1984) *Normal Accidents.* Perrow begins his text by outlining, and not entirely dismissing, 'conventional' explanations for accidents – not least 'operator error', lack of attention to safety features, lack of operating experience, and so on – but goes on to argue that there are more basic and important contributory factors to the failures of systems (ibid.: 63). This more fundamental phenomenon is the very nature of the processes and organisation that characterise some forms of production.

Thus Perrow claims that, for example, chemical plants are based on complex rather than linear reactions, the former being interactions 'in which one component can interact with one or more other components outside of the normal production sequence' (ibid.: 77–8). Indeed, complex systems are increasingly the norm in certain industries given their 'advantages' over linear systems:

> Complex systems are more efficient (in the narrow terms of production efficiency, which neglects accident hazards) than linear systems. There is less slack, less underutilised space, less tolerance of low quality performance, and more multifunction components. From this point of view, for design and hardware efficiency, complexity is desirable. (ibid.: 88)

Moreover, according to Perrow, some industries which are based on complex rather than linear reactions are also characterised by 'tight' rather than 'loose' coupling. Tight coupling suggests that: delays in processing are not possible; little slack is possible in supplies, equipment and personnel; there are limited possibilities of substitution of these resources; and there is an absence of designed-in buffers and redundancies (ibid.: 96). In other words, the nature of disasters and 'accidents' (for Perrow), and the nature of safety crimes (for us), is misunderstood by remaining at the level of the individual; one must focus upon the nature of the system of which the individual is a part.

The actions, inactions, decisions, and so on of individuals must therefore be placed in the structures within which they operate – and this means taking cognisance of their immediate workgroup, their workplace, their organisation/company as well as, beyond these, a far wider complex of factors. This highlights the need to move beyond individualised or environmental explanations, to engage in the theoretical and empirical work required to develop an understanding of safety crimes. And the urgency of developing such a wide-ranging explanatory framework has been raised by some commentators on corporate crime, who have emphasised the need to incorporate explanatory variables which range from the micro (individual) through to the macro (socio-structural) levels (Coleman 1987; Punch 2000; Vaughan 1992, 1996). As with theory-building in corporate crime research in general (Cressey 1989), theoretical development here remains at an early stage, although there are now a number of book-length studies which attempt to use some of this range of factors. These include studies of safety crimes in the offshore oil industry (Whyte 1999a, Woolfson *et al.* 1996), corporate crime in the asbestos (Tweedale 2000), chemicals (Pearce and Tombs 1998) and pharmaceutical (Braithwaite 1984) industries, corporate manslaughter (Slapper 2000), corporate and state illegalities associated with the fateful launch of the Challenger space shuttle (Vaughan 1996), and a national case study in the micro/macro control of 'economic' crime (Alvesalo 2003b). All of those latter studies integrate analyses of systematic features in the production of corporate crimes. The same features can be identified in the case studies that we set out in Chapter 1. Here we organise those features into three levels.

First, at the micro-level of the individual and of inter-personal relations, we need to take account of individual personality and characteristics, not least in terms of the kinds of personalities that are recruited or encouraged within the organisation, as well as 'individual' factors that are socially constructed as relevant such as rank/position within hierarchy, age, gender, and ethnicity. Shifting to the level of the immediate work-group or sub-unit within the organisation, we must take account of inter-personal dynamics (and particularly the possibility of 'groupthink'), the culture of the work-group (and the extent to which this coheres or clashes with the culture of the wider organisation), and its location within the overall organisation, both structurally and geographically – that is, is it relatively autonomous or highly supervised? Is it part of one large organisational complex, or is it geographically isolated?

Second, at what might be called the meso-level, there are also key sets of issues to be raised in relation to the organisation itself. At this level, we need to understand something of its organisational structure, its internal lines of decision-making and accountability, its geographical scope of operations, and the nature, volume and complexity of internal transactions. Issues of organisational culture must also be addressed: is the organisation risk taking or risk averse; it is gendered; is it authoritarian; and is it one where a blame culture predominates? Crucially, we must also enquire into what kind of management is either valorised or deemed acceptable by the organisation; and related to this point are issues concerning workforce organisation, notably the existence, and strength, both of trade unions, as well as of safety reps and safety committees. Of further relevance are the very products or services that are the focus of the organisation: are these opaque or transparent, are they sold to consumers or other organisations, is their production labour intensive or capital intensive? Perhaps most obviously, we need to know something of the economic 'health' of the company, and of specific organisational units, as well as the ways in which, and the time-scales across which, profitability is calculated.

Finally, there are key sets of questions to be broached regarding the macro-economic, political and social environments within which the organisation operates. Amongst these extra-organisational features are: the nature of the market structure; the size and scope of the market; the predominant form of inter-organisational relationships within any given market; the material and ideological state of regulation; the more general nature of state–business relationships; the dominant form of political economy, and concomitant societal values, including the nature and degree of pro- or anti-business sentiment. Thus an understanding of changing levels of occupational 'safety' requires attention to factors which include the following: the abilities of capital to relocate within and beyond national boundaries, and thereby to export hazard and risk; trends in labour markets and employment patterns, not least attempts to imitate models of flexibility claimed to exist in the US and Japan; the nature of, and changes in, contractual arrangements and methods of payment; and the introduction of new technologies and new forms of work organisation, particularly in terms of the impact of these upon workers' skills and functions. Crucially, as the case studies on Piper Alpha, Bhopal, Morecambe Bay and the *Herald of Free Enterprise* illustrate most clearly, we need to gain an understanding of the national and international economic conditions that shape organisational strategies.

Whether examined in isolation, or in their combination through attempts to conceptualise the range and relative importance of such factors in terms of an overarching framework, the production of safety crimes therefore needs to be conceptualised at a range of micro, meso and macro levels. And this requirement, for an integrated understanding of these complex levels of analysis, takes us beyond criminology. Indeed, even in our brief discussion of classicisms, and individual and sociological positivisms, it is clear that we needed to move towards disciplinary areas such as organisation theory and organisation studies, economics, and political science in order to fully understand how corporate crimes are produced in any given society. This task in fact requires a political economy (Gill and Law, 1988: xviii) of safety crimes, one which understands their production through prevailing systems of economic, social and political organisation, dominant value systems and beliefs, and the differential distribution of power. Further, the integrated historical and international focus of political economy forces us to recognise that we cannot understand safety crimes in the UK without some understanding of how these, first, have emerged, and, second, fit within broader market processes that are increasingly played out globally. If we return to the case of Simon Jones, we see just how vital a broader political economy is to glean a full understanding of the production of safety crimes. Thus, we find that just as changes in the labour market summarised under the label casualisation explain why he was able to work in a high risk industry with neither training nor supervision, we need to turn to an analysis of a welfare state under reconstruction to understand what led him to be working at all. Remember that Simon Jones was a student, taking a year out before completing his degree programme. He had signed up with Personnel

Selection, in the words of his friend, to 'get the dole off his back'. His being in the place where he could be killed was partly the result of a benefit system which, under the guise of encouraging people to work, penalises them for failing to do so, particularly in the case of young people. For some, these may be taken-for-granted facts of life. But it was not always so, and indeed the state of affairs is both recent and can only be understood in the context of macro-level social processes, on both national and international levels.

A thematic political economy of safety crimes

Understanding the complexities of safety crimes therefore means that we need to move beyond criminological theory and take account of the micro, meso and macro processes that we noted in the previous section. One way to develop a systematic analysis of safety crimes is to formulate a more thematic analysis of the material that we have discussed in this book. This is the point at which we can begin to make more sense of the themes that were sketched out in Table 1, in Chapter 1, and to ask, in particular, how do they help us to understand the incidence of (and various reactions to) safety crimes? To this end, we can identify six key features in the production of safety crimes that emerge from our analysis of safety crimes throughout this text and elaborate upon our initial analysis in Chapter 1.

1. Worker's vulnerability to victimisation

The cases raised various factors related to the nature of workforces, which may also be viewed as the *objects* of managerial practices and cultures. When grouped together, we find a key theme is that workers at plants or in sectors where safety crimes occur are disproportionately likely to be casualised, sub-contracted and/or increasingly migrant workforces. This is not to claim that these are the *only* workers who are victims of safety crimes – victimisation by such crimes is ubiquitous – but as we saw in Chapter 2, for example, *vulnerability* is often a characteristic of those who are victimised by such crimes. More generally, vulnerability in relation to death and injury, let alone safety crimes *per se*, is a function of the strength of workers in relation to managements. As we also noted in Chapter 2, research evidence overwhelmingly points to the finding that the safest workplaces are those workplaces that have strong trade union representation (James and Walters 2002; Reilly *et al.* 1995; Walters *et al.* 2005) suggesting that the relative strength of workers in the workplace is the key factor in the amelioration of the risk of safety crimes.

2. *The nature of management*

First, there are features about the companies themselves, in terms of how they are managed, how workers are treated, and the standards of management that can be regarded as 'acceptable'. Here we find numerous references to aggressive managements, and to managements who ensured that warnings, usually from below, were being systematically ignored. Moreover, we identify a generalised lack of management accountability for safety crimes. Management decisions and failures to heed warnings are subject to very little external counter-balancing in terms of regulation. Indeed, through processes of victim blaming, those providing warnings – and who were ultimately to pay a high price for these being ignored – are often cast as the problem itself. As we saw in Chapter 3, these claims both produce and are in turn supported by generalised discourses of victim blaming via the notion of the 'accident-prone' worker. Moreover, as we noted, this victim blaming can have long-term consequences, leading, for example, to a lack of compensation for victims. Further, we find managements who design or maintain plant badly – that is, managers who do not manage very well. And this is further reflected by the fact that among the cases in Chapter 1 we find several instances not just of safety crimes but of environmental offences, and of companies that frequently offend or sectors where recidivist employers appear to be commonplace. Thus managerial practices – and the cultures within which these are embedded – are crucial in understanding the production of safety crimes.

3. *Inter-organisational factors*

If the relationship between workers and managers is the key factor in determining the risk of safety crimes being produced, then this is also related to features in the labour market that enable work to be sub-contracted or subdivided between organisations. There is strong evidence that points to a weakening of workers' position in relation to the labour market through casualisation and sub-contracting and their vulnerability to workplace risks (Mayhew and Quinlan 1998; Mayhew *et al.* 1997). That is, workplace balances of power are related to, though not simply determined by, wider social relations of power. Clearly, this needs to be understood in an extra-organisational context. Beyond the organisational structures and cultures of companies, there are key *inter*-organisational features which seem to emerge consistently. These include the ways in which different parts of the same firm relate to each other (parent–subsidiary relationships, for example, in the case of Bhopal, and agency–contractor relationships, in the case of Simon Jones and Piper Alpha), how different firms were linked to each other within or across particular sectors, perhaps in terms of long and complex supply chains, or indeed in terms of systematic relationships between legal and illegal businesses (in the case of Morcambe Bay).

4. Market factors

A further group of factors, both cultural and organisational, is located at the level of the market or industry in question. These include the norms that predominate in an industry for what is acceptable or even 'required' in terms of how production of goods and services is organised – characterised, for example, in the 'political economy of speed', alongside a series of (more or less real, but perceived in any case) market pressures, operating locally, nationally, and internationally, or even globally. Thus in *each* of the cases examined in Chapter 1, particular features of a market create quite specific demands for profitability, speed or cost-cutting. This location of the problem of workplace safety in the market has been developed by Moore (1991). He examined the creeping casualisation of the organisation of work in a number of industries and, in the context of a deregulated economy and the reassertion of the supremacy of the market, describes the result as 'cash-nexus carnage'. As others did before him did (for example, Nichols and Armstrong 1973; Carson 1980c), Moore sees the intensification of increasingly unfettered markets as the greatest threat to the health and safety of workers, thus offering the combination of opportunity plus motivation for safety crimes.

5. Law and regulation

A key group of factors to be considered in any explanation of safety crimes is located in law and regulation. Occupational safety is regulated in a way that, as we found systematically through our discussion in this text (particularly in Chapters 6, 7 and 8), there are consistent failings in or problems with legal categories and reasoning, ongoing regulatory oversight, and in post-event legal and regulatory processes. But even before we can understand contemporary law and enforcement with regard to occupational safety, we need to recognise that the bases of these are to be found in the ways in which legal systems have been constructed either to separate out safety crimes from real crimes, or even to deny the notion of safety crimes. Thus, low levels of inspection, detection, formal enforcement and sanctions ensure that safety offences are regarded as less serious than other crimes of violence, an enduring phenomenon that acts to reduce the social opprobrium that is attached to those crimes. Crucially, the way the state responds to safety crimes shapes the way that those crimes are tolerated from the boardroom to the workplace.

6. Unequal distribution of power

This takes us, finally, to a common theme of this text, namely power, one which clearly links our discussions on the production of safety crimes, their representation, and their regulation, but which also can be seen as a particular quality that accrues to businesses in relation to states, other organisations and populations. Thus we found in the cases in Chapter 1 and the discussion of 'structured vulnerability' in Chapter 2, how changes in social policy through the structure of the welfare state, as well as changes in state forms and public policy more generally, have augmented the leverage enjoyed over populations and states by private capital. This may be seen locally, as in local and/or national state support for investment by an offending company in a particular locale, or in the conflict this appears to raise between safe jobs or no jobs at all, or more generally in the power of business to circumvent, or prevent resort to, legal processes. And, finally, and to reassert the international dimension of the phenomena with which we are dealing in this text, the significance of the power and reach of multinational corporations, and their ability to trade off 'developed' versus 'developing' economy safety standards, illustrates the insidious ways in which unequal power relations can act to encourage safety crimes.

Conclusion

The constituent themes of a political economy of safety crimes, sketched above, do not fit well with the way that criminology and law constructs these phenomena. If we have already discussed previously, and in this chapter, various aspects of the ways in which criminology acts to obscure safety crimes, we have also pointed in Chapters 5–8 to the ways in which safety crimes are constructed through law and its application to them. One feature that unites both criminology and law in this respect is that they both reject political economy explanations in favour of a highly individualised explanation of 'accidents' at work, which separates out the phenomena with which we are concerned from 'real' crimes. To illustrate, as we saw in Chapter 7, health and safety law is underpinned by a philosophy that asserts a natural identity of interests between workers and management in the organisation of workplace safety. This idea is enshrined in the Robens Report which formed the basis of the Health and Safety at Work Act (1974). Particularly important points are made in this respect about Robens's conviction that 'safety is mainly a matter of the day to day attitudes and reactions of the individual' (Robens Report: para 13). Robens had argued that the key problem in effective safety organisation was 'worker apathy': the reluctance of workers to play an active role in preventing 'accidents'. Indeed, he argued that if safety law was structured in a rigid and prescriptive way, and if state regulators adopted strict enforcement policies, then this might discourage workers from being involved in taking appropriate action to prevent 'accidents'. In other words, for Robens, stricter enforcement might actually encourage rather than prevent safety crimes. Anthony Woolf, however, pointed out in his critique of the Robens Report that the weakness of the 'natural identity of interests' philosophy is principally located in its inability to grasp that 'nearly all accidents are the inevitable result of unsafe working systems which could themselves be made safe by the employers, by a combination of hazard analysis, planning, training and supervision' (1973, ibid.: 90). Woolf questioned Robens's assumptions on apathy, arguing that: 'if they are false, as I believe them to be, the role envisaged by Robens for law and order enforcement will itself have to be examined' (ibid. 89). If this text has demonstrated one thing, then it is that, more than 30 years after Woolf's assessment of Robens, enforcement around safety crimes remains wholly inadequate.

A fundamental question that has not been stated, but is at least implicit throughout much of this book, is whether, given the manifest failure of the criminal justice system to prevent crime (Hillyard and Tombs 2004), we should or could expect things to be any different when it comes to safety crime. But for us this is perhaps the wrong way of posing a potentially significant question. For, although the State has indeed failed, particularly in relation to under-enforcement, to talk only of State failure in the face of rising crime rates encourages us to think of the State in a particular way: that the State has lost control *over* society. This may be true, but a more useful way of thinking about State power is how it operates *through* society. The State is not simply an open system, vulnerable only to the demands of the most powerful or effective lobbying by interests groups. States in capitalist social orders are precisely that: capitalist states. States pursue policies that privilege the interests of capital over labour, reinforce unequal gender relations, and reproduce racist and other forms of social divisions. But, although social divisions are manifestly reinforced and exacerbated by the State, State institutions, as we noted in Chapter 7, also remain (unequal) structures of representation. The stability of the social order is only guaranteed after struggles between different groups are absorbed by the State. This involves not simply a process of ruling over sub-dominant groups, but also a compromise that secures the legitimacy of the rulers and consent of the ruled. One consequence of this process is that State institutions remain vulnerable to counter-hegemonic demands. If the recent history of capitalist societies has taught us anything, it is that the struggle against State–corporate power is most effectively resisted by collective opposition from workers and from other popular movements in workplaces, at sites of consumption and on the streets. So whilst there is a case for bringing the *State* back into our understanding of how best to oppose power, we must at the same time recognise the pivotal position that workers and the public have in struggles against safety crimes (Tucker 1990).

Both the under-enforcement of regulation and the absence of controls for serious corporate harm provides us with a perfect illustration of Lacey's (1994) point that we cannot take 'crime' for granted. Neither can we take its enforcement for granted. The criminal justice system remains pre-occupied with a relatively limited number of interpersonal crimes. With the key exception of the so-called 'war on terror', Home Office priorities have been, and remain, a limited number of 'headline' crimes of interpersonal violence, property and public 'nuisance'. And criminology by and large obediently falls into line behind this 'official' version of what constitutes the crime problem (Hillyard *et al.* 2004; Walters 2003). The law's embedded bias, which – as we saw in Chapter 5 – is historically inscribed, forces us to think about a much more profound question in relation to regulation: whether under capitalist social orders, State regulation can ever adequately guarantee our protection? Can workplace deaths and injuries ever be sufficiently harnessed in states that see as their primary aim the encouragement of private profit maximisation and capital growth? Given that under-regulation and an absence of safety crime controls appears to be as much an embedded feature of capitalist social orders as safety *crime* itself is, our greatest challenge is perhaps not simply to reform or tinker with the means we have to control individual corporations. A much greater and more pressing challenge is to seek an alternative means of organising production regimes that will neither encourage nor sustain the routine killings and injuries committed by employers.

Notes

1 For the HSE Better Business strategy, see www.hse.gov.uk/betterbusiness/large/index.htm.

2 One has only to watch *Wall Street* to see that the personality traits that helped to make Gecko a success were the same characteristics that allowed him, through a combination of intention and negligence, to break the law.

3 This is also one example of 'techniques of neutralisation', which are crucial to understanding motivation in the context of corporate crime: see Box 1983; Slapper and Tombs 1999.

Sources of further information

Almost all of the key texts – books, chapters, journal articles, reports – which deal with safety either from a criminological or social science perspective have been cited in this book. We will not rehearse these in detail here, save to point again to a few key items. Our discussion of these follows the structure of the book.

Numerous case studies of safety crimes exist – though these are overwhelmingly written from the US and thus the selection of such cases is also biased either towards their occurrence in the US or to US-based companies. This small caveat aside, most collections of case studies of corporate crimes contain some examples of safety crimes, but see in particular a useful selection of cases across various chapters of Hills (1987b), Pearce and Snider (1995) and Tucker (2006).

The relative invisibility of safety offences has been considered, in terms of the various elements of this legal, political and social process, in Bergman (1991), Glasbeek and Rowland (1979), Glasbeek (1989), McMullan and McClung (2006), Tombs (1999b, 2000) and Tombs and Whyte (2001). Two classic texts in which safety crimes are considered – however briefly – in the context of an excellent general critique of the biases of criminal justice systems and representations of the crime problem towards the crimes of the relatively powerless are Box (1983) and Reiman (1998). Efforts to document the *actual* scale of death, injury and safety crimes are very rare – not least because of the problems of access to data. For this reason, despite it now being a *little* dated, the work undertaken by David Bergman and the CCA remains seminal here (see, in particular, Bergman 1993, 1994 and 2000; and Unison/CCA 2002). A more forensic approach to a much smaller selection of cases (all involving legal responses to work-related death) can be found in Slapper (2000), a key text for understanding the failures of the criminal justice system to treat safety crimes *as crimes*.

The classic texts on the emergence of law with respect to health and safety are undoubtedly those by Carson (1970a, 1970b, 1974, 1979, 1980a and 1980b); see also the critique offered by Bartrip and Fenn (1980a, 1980b, and 1983). For an historical discussion that bears similarity in approach to that of Carson, albeit in a Canadian context, see Tucker (1990). An excellent general context for understanding the nature and historical emergence of contemporary law and legal systems can also be found in Norrie (2001).

On regulation, a debate often referred to is that around the relative strengths and merits of a 'compliance-oriented' versus a more punitive regulatory approach – see Hawkins (1990 and 1991), and Pearce and Tombs (1990 and 1991). For critical, and more recent, commentary on the debate, see Davis (2004), Hopkins (1994), Johnstone (1999) and Gray (2006). On the regulatory mix, the work of Braithwaite and colleagues is currently dominant: see Ayres and Braithwaite (1992), Gunningham and Johnstone (1999), Haines (2005) and Johnstone (2003). For a typology of approaches to regulation, see Whyte (2004). Hawkins (2002) is a fascinating study, based upon rare insider access, of the decision-making processes employed by safety inspectors in the UK as to whether or not to prosecute for safety offences. The classic piece on regulatory reform – including a specific discussion of safety regulation – remains Snider (1991). On the sanctioning of safety offenders, the most useful – and still very recent – overviews are set out by Croall (2005) and the CCA.[1]

A general political economy of occupational safety is set out in Nichols (1997), though this is not organised around safety crimes *per se*. Attempts to account for the production of safety crimes in an integrated fashion – encompassing the various levels and elements of analysis that we have argued must be part of any adequate account – can be found in Carson (1982), Davis (2000), Whyte (1999a), Woolfson et al. (1996), Pearce and Tombs (1998), Slapper (2000), Vaughan (1996), and, though couched somewhat differently, an excellent and concise discussion of the wide range of factors to be taken into account in any causal explanation of safety crimes is set out in Punch (2000).

Beyond these texts, journals and book chapters, for us the key source of critical information on the state of British safety – and one which frequently discusses safety crimes in particular – has long been, and remains, *Hazards* magazine. *Hazards* aims to look behind the company safety hype, and provide practical tools for workplace safety representatives to prevent safety crimes. Using a global network of union safety correspondents, *Hazards* provides the best first port of call for information on safety and safety crimes. It can be accessed at www.hazards.org/.

More generally, there is, of course, a great deal of information available electronically. There are few, if any, websites devoted to safety crimes *per se*, but there are, first, many devoted to issues of safety and, second, to corporate crime in general (among the latter, those with the best coverage of safety crimes are listed).

On 'safety' and safety crimes, see the following:

www.hazardscampaign.org.uk

The Hazards Campaign, a network of resource centres and campaigners on health and safety at work, and the best source of safety contacts and campaigning groups information in the UK.

www.corporateaccountability.org

The Centre for Corporate Accountability, a charity concerned with the promotion of worker and public safety, provides free advice to victims of safety crimes, campaigns around law and enforcement, and undertakes related research, much of which is available at the site.

www.tuc.org.uk/h_and_s/index.cfm

The health and safety section of the Trades Union Congress website. Note, in particular, the newsletter *Risks*, which can be accessed from this site.

www.worksmart.org.uk/rights/viewsection.php?sen=9

Advice on health and safety law provided to workers by the Trades Union Congress.

www.labournet.net/default.asp

LabourNet aims to be in the forefront of using the resources of the internet to provide communications, news and information for the labour movement, seeking to strengthen organised labour. There are excellent sections on safety.

www.iosh.co.uk

Website of the Institution of Occupational Safety and Health which is the UK professional body for health and safety practitioners. The Institution of Occupational Safety and Health's stated aims are to regulate and steer the profession and provide members and groups outside the organisation with guidance on health and safety issues.

For sites devoted either specifically, or with useful material on, corporate crime in general, and with some useful coverage of safety crimes, see the following:

www.corpwatch.org

Corporate Watch (US) tracks illegal and unethical corporate activity, and business–industry relations. Its parent organisation is the Transnational Resource and Action Center (TRAC), based in San Francisco.

www.corporatewatch.org

Corporate Watch (UK) tracks similar forms of corporate activity to the US journal of the same name (above), but is independent of and unrelated to the US publication.

www.essential.org

Multinational Monitor, published by Essential Information, Inc., tracks corporate activity, especially in the 'Third World', focusing on the export of hazardous substances, worker health and safety, labour union issues and the environment.

www.corporatepredators.org

Here you will find 'Focus on the Corporation', a weekly column on illegal and unethical corporate activity, posted by two North American journalists/activists.

http://paulsjusticepage.com/elite-deviance.htm

An excellent source of corporate crime material, and many links, maintained by Paul Leighton, co-author of Reiman's *The Rich get Richer and the Poor Get Prison.*

www.nader.org

The Nader Page. The site of Ralph Nader, long time US anti-corporate campaigner, this site seeks to further the ability of consumers to be heard and to have a real voice and a significant role in the legislative and regulatory decision-making processes.

www.motherjones.com

Mother Jones is an independent not-for-profit US-based campaigning site and magazine whose roots lie in a commitment to social justice implemented through first rate investigative reporting.

www.citizen.org

Public Citizen, an organisation that campaigns for openness and democratic accountability in government; for the right of consumers to seek redress in the courts; for clean, safe and sustainable energy sources; for social and economic justice in trade policies; for strong health, safety and environmental protections; and for safe, effective and affordable prescription drugs and health care.

'Official' information on health and safety can be found at the following sites, each of which also has plenty of useful links:

www.hse.gov.uk

The Health and Safety Executive, contains an increasingly useful body of information; see especially sections on Enforcement, the (searchable) Notices and Prosecutions database, Statistics, Annual Reports, Local Authorities, and Press Releases.

http://uk.osha.eu.int

European Agency for Safety and Health at Work, established by the EU to collect, analyse and promote OSH-related information, with the stated aim of making Europe's workplaces safer, healthier and more productive, and in particular to promote an effective prevention culture.

www.osha.gov

Occupational Safety and Health Administration, US equivalent to HSE, its stated aim is to prevent work-related injuries, illnesses and deaths for every American worker.

www.cdc.gov/niosh/homepage.html

The National Institute for Occupational Safety and Health, a US federal agency responsible for conducting research and making recommendations to the US government on work-related injury and illness.

www.ilo.org

The International Labour Organization, a UN agency, sets, promotes and supervises international labour standards, and is the UN's main source of expertise about workers' rights and conditions, social protection and labour law.

Note

1 See, in particular, 'Sentencing Issues' briefing at: www.corporateaccountability.org/rb/sentencing/sentencing.htm, downloaded, December 3, 2006.

References

Alvesalo, A. (2003a) 'Economic Crime Investigators at Work', *Policing and Society*, 13(2): 115–38.

Alvesalo, A. (2003b) *The Dynamics of Economic Crime Control*, Espoo: Poliisiammattikorkeakoulun tutkimuksia.

Alvesalo, A. and Tombs, S. (2001) 'The Emergence of a "War" on Economic Crime: The Case of Finland', *Business and Politics*, 3(3): 239–67.

Amnesty International (2004) *Clouds of Injustice. Bhopal disaster twenty years on*, Oxford: Amnesty International, at web.amnesty.org/wire/December2004/Bhopal, downloaded, December 2, 2006.

Anderson, B. and Rogaly, B. (2005) *Forced Labour and Migration to the UK*, London: COMAS/TUC.

Australian Broadcasting Company (2004) Protests held over death of Olympic construction workers – Wednesday, August 11. www.abc.net.au/sport/content/200408/s1173882.htm, downloaded, December 2, 2006.

Ayres, I. and Braithwaite, J. (1992) *Responsive Regulation: Transcending the Deregulation Debate*, Oxford: Oxford University Press.

Babiak, P. and Hare, R.D. (2006) *Snakes in Suits. When Psychopaths Go To Work*, New York: HarperCollins.

Bakan, J. (2004) *The Corporation. The Pathological Pursuit of Profit and Power*, New York: Free Press.

Baldwin, R. (1995) *Rules and Government*, Oxford: Clarendon Press.

Baldwin, R. (1997) 'Regulation: After "command and control"', in K. Hawkins (ed.) *The Human Face of Law*, Oxford: Clarendon Press: 65–84.

Baldwin, R. and Cave, M. (1999) *Understanding Regulation: Theory, Strategy and Practice*, Oxford: Oxford University Press.

Baldwin, R., Scott, C. and Hood, C. (eds) (1998) *A Reader on Regulation*, Oxford: Oxford University Press.

Bardach, E. and Kagan, R. (1982) *Going by the Book: The Problem of Regulatory Unreasonableness*, Philadelphia: Temple University Press.

Bartrip, P. and Fenn, P.T. (1980a) 'The Conventionalization of Factory Crime – A Reassessment', *The International Journal of the Sociology of Law*, 8: 175–86.

Bartrip, P. and Fenn, P.T. (1980b) 'The Administration of Safety: The Enforcement Policy of the Early Factory Inspectorate 1844–1864', *Public Administration*, 58: 87–102.

Bartrip, P. and Fenn, P.T. (1983) 'The Evolution of Regulatory Style in the Nineteenth Century British Factory Inspectorate', *Journal of Law and Society*, 10(2): 201–22.

BBC Radio 4 (2004) *You and Yours – Transcript*, Friday, July 23, www.bbc.co.uk/radio4/youandyours/yy_ftf_athens_230704.shtml, downloaded, December 2, 2006.

Beck, M. and Woolfson, C. (1999) 'Safety Culture – A Concept Too Many?', *The Health and Safety Practitioner*, 16(1): 14–16.

Becker, G. (1968) 'Crime and Punishment: An Economic Approach', *Journal of Pol.itical Ecomony*, 76(2): 169–217

Bergman, D. (1988) 'The Sabotage Theory and the Legal Strategy of Union Carbide', *New Law Journal*, 138, June.

Bergman, D. (1991) *Deaths at Work. Accidents or Corporate Crime*, London: Workers' Educational Association.

Bergman, D. (1992) 'Corporate Sanctions and Corporate Probation', *New Law Journal*, 142, September.

Bergman, D. (1993) *Disasters. Where the Law Fails. A New Agenda for Dealing with Corporate Violence*, London: Herald Families Association.

Bergman, D. (1994) *The Perfect Crime? How Companies Can Get Away with Manslaughter in the Workplace*, Birmingham: West Midlands Health and Safety Advice Centre.

Bergman, D. (2000) *The Case for Corporate Responsibility: Corporate Violence and the Criminal Justice System*, London: Disaster Action.

Bergman, D. and Fooks, G. (forthcoming) *Sentencing Health and Safety Offences: A Report for the Health and Safety Executive*, London: HSE.

Bernstein, M. (1955) *Regulating Business by Independent Commission*, Princeton, New Jersey: Princeton University Press.

Bhopal Action Group (1988) *Sabotaging the Sabotage Theory: A Critique of the Paper by Ashok Kalelkar*, London: Transnational Information Centre.

Bianchi, H. and van Swaaningen, R., (eds) (1986) *Abolitionism. Towards a Non-Repressive Approach to Crime*, Amsterdam: Free University Press.

Bibbings, R. (1996) *Managing Occupational Road Risk. Discussion Paper*, Birmingham: RoSPA.

Black, J. (1996) '"Which Arrow?": Rule Type and Regulatory Policy', in D. Galligan (ed.) *A Reader on Administrative Law*, Oxford: Oxford University Press: 165–93.

Black, J. (1997) *Rules and Regulators*, Oxford: Clarendon Press.

Boris, E. and Pruegl, E., (eds) (1996) *Homeworkers in Global Perspective. Invisible No More*, London: Routledge.

Bourdieu, P. (1998) *Acts of Resistance: Against the Tyranny of the Market*, New York: The New Review Press.

Bowie, V. (2002) *Workplace Violence*, New South Wales: Workcover.

Box, S. (1983) *Power, Crime and Mystification*, London: Tavistock.

Boyd, C. (1992) 'The Zeebrugge Car Ferry Disaster', in W. Frederick, J. Post and K. Davis, *Business and Society. Seventh Edition*, Hightstown, NJ: McGraw Hill: 498–511.

Braithwaite, J. (1982) 'Enforced Self Regulation: A New Strategy for Corporate Crime Control', *Michigan Law Review*, 80: 1466–1507.

Braithwaite, J. (1984) *Corporate Crime in the Pharmaceutical Industry*, London: Routledge and Kegan Paul.

Braithwaite, J. (1989) *Crime, Shame and Reintegration*, Cambridge: Cambridge University Press.

Braithwaite, J. (1993) 'Responsive Regulation for Australia', in P. Grabosky, and J. Braithwaite (eds) *Business Regulation and Australia's Future*, Canberra: Australian Institute of Criminology: 81–96

Braithwaite, J. (2000) *Regulation, Crime, Freedom*, Aldershot: Ashgate.

Braithwaite, J. (2002) 'Rewards and Regulation', *Journal of Law and Society*, 29(1): 12–26.

Braithwaite, J. and Fisse, B. (1987) 'Self Regulation and the Control of Corporate Crime', in C. Shearing, and P. Stenning (eds), *Private Policing*, Beverly Hills: Sage: 221–246.

Braithwaite, J. and Geis, G. (1982) 'On Theory and Action for Corporate Crime Control', *Crime and Delinquency*, 28(2): 292–314.

Brake, M. and Hale, C. (1992) *Public Order and Private Lives: The Politics of Law and Order*, London: Routledge.

Brooks, L. (2001) 'Alarm as employer cleared in death case', the *Guardian*, November 30.

Bryman, A. (ed) (1988) *Doing Research in Organisations*, London: Routledge.

Calavita, K. (1983) 'The Demise of the Occupational Safety and Health Administration: A Case Study in Symbolic Action', *Social Problems*, 30(4): 437–8.

Campbell, D. (2005) 'Working drivers responsible for 1,000 road deaths a year', *Observer*, November 13.
Carrabine, E., Iganski, P., Lee, M., Plummer, K. and South, N. (2004) *Criminology. A Sociological Introduction*, London: Routledge.
Carson, W.G., (1970a) 'White-Collar Crime and the Enforcement of Factory Legislation', *British Journal of Criminology*, 10: 383–98.
Carson, W.G. (1970b) 'Some Sociological Aspects of Strict Liability and the Enforcement of Factory Legislation', *Modern Law Review*, 33 (July): 396–412.
Carson, W.G. (1974) 'Symbolic and Instrumental Dimensions of Early Factory Legislation: A Case Study in the Social Origins of Criminal Law', in R. Hood (ed.) *Crime, Criminology and Public Policy*, London: Heinemann: 107–138.
Carson, W.G. (1979) 'The Conventionalisation of Early Factory Crime', *International Journal of the Sociology of Law*, 7(1): 37–60.
Carson, W.G. (1980a) 'Early Factory Inspectors and the Viable Class Society – A Rejoinder', *International Journal of the Sociology of Law*, 8(2): 187–91.
Carson, W.G. (1980b) 'The Institutionalisation of Ambiguity: Early British Factory Acts', in G. Geis and E. Stotland (eds.) *White-Collar Crime. Theory and Research*, London: Sage: 142–173.
Carson, W.G. (1980c) 'The Other Price of Britain's Oil: Regulating Safety on Offshore Oil Installations in the British Sector of the North Sea', *Contemporary Crises*, 4: 239–66.
Carson, W.G. (1981) 'White-Collar Crime and the Institutionalisation of Ambiguity: The Case of the Factory Acts', in M. Fitzgerald, G. McLennan, and J. Pawson, (eds) *Crime & Society. Readings in history and theory*, London: RKP/OUP.
Carson, W.G. (1982) *The Other Price of Britain's Oil: Safety and Control in the North Sea*, Oxford: Martin Robertson.
Cassells, J. (1993) *The Uncertain Promise of Law. Lessons from Bhopal*, Toronto: University of Toronto Press.
Castleman, B. (1979) 'The Export of Hazard to Developing Countries', *International Journal of Health Services*, 9(4): 569–606.
Catley, B. (2003) *Philosophy – the Luxurious Supplement of Violence*, paper presented at Critical Management Studies 2003, Lancaster, July, available at www.mngt.waikato.ac.nz/ejrot/cmsconference/2003/proceedings/philosophy/catley.pdf, downloaded December 2, 2006.
Centre for Corporate Accountability (1999) *Response to the Consultation of the Sentencing Advisory Panel, November 1999*, London: CCA.
Centre for Corporate Accountability (2004) *Evidence to the House of Commons Work and Pensions Committee, The Work of the HSC/E*, London: CCA.

Chambliss, W. (1967) 'Types of Deviance and Effectiveness of Legal Sanction', *Wisconsin Law Review*, 3, Summer, 703–19.

Chibnall, S. (1977) *Law-and-Order News. An Analysis of Crime Reporting in the British Press*, London: Tavistock.

Chomsky, N. (1999) *Profit Over People: Neo-Liberalism and Global Order*, New York: Seven Stories Press.

Clarke, M. (2000) *Regulation. The Social Control of Business Between Law and Politics*, London: Macmillan.

Clarkson, C. and Keating, H. (1994) *Criminal Law: Text and Materials*, 3rd edn, London: Sweet and Maxwell.

Clinard, M. and Yeagar, P. (1980) *Corporate Crime*, New York: The Free Press.

Clive, E., Ferguson, P., Gane, C. and McCall Smith, A. (2003) *A Draft Criminal Code for Scotland with Commentary*, Edinburgh: Scottish Law Commission.

Coffee, J. (1981) 'No Soul to Damn, No Body to Kick: An Unscandalised Inquiry Into the Problem of Corporate Punishment', *Michigan Law Review*, 79(3): 386–459.

Cohen, S. (1985) *Visions of Social Control*, Cambridge: Polity.

Coleman, C. and Moynihan, J. (1996) *Understanding Crime Data*, Buckingham: Open University Press.

Coleman, J.S. (1987) 'Toward an Integrated Theory of White-Collar Crime', *American Journal of Sociology*, 93(2): 406–39.

Construction Skills Network (2006) *Blueprint for UK construction skills 2006–2010*, www.constructionskills.net/research/constructionskill-snetwork/forecastmodel/constructionskillsnetworkoutputs/uk/, downloaded December 2, 2006.

Conscious Cinema (2000) *Not This Time – The Story of the Simon Jones Memorial Campaign*, film available at www.simonjones.org.uk.

Cotton, J. (2004) 'Homicide', in D. Povey, (ed), *Crime in England and Wales, 2002/2003. Supplementary Volume 1. Homocide and gun crime*, 01/04, London: Home Office.

Craig, M. (1981) *Office Workers' Survival Handbook. A Guide to Fighting Health Hazards in the Office*, London: The Women's Press.

Crainer, S. (1993) *Zeebrugge. Learning From Disaster*, London: Herald families' Association.

Cressey, D. (1989) 'The Poverty of Theory in Corporate Crime Research', in F. Adler and W.S Laufer (eds), *Advances in Criminological Theory*, New Brunswick, NJ: Translation, 31–55.

Croall, H. (2001) 'The Victims of White-Collar Crime', in S-Å. Lindgren (ed.), *White-Collar Crime Research. Old Views and Future Potentials. Lectures and Papers from a Scandinavian Seminar* (BRÅ-Rapport 2001:1), Stockholm: Brottsförebyggande råde/Fritzes.

Croall, H. (2005) 'Penalties for Corporate Homicide', annex to *Scottish Executive Expert Group Report, Corporate Homicide*, Edinburgh: Scottish Executive.

Croall, H. and Ross, J. (2002) 'Sentencing the Corporate Offender: Legal and Social Issues', in N. Hutton and C. Tata (eds) *Sentencing and Society: International Perspectives*, Aldershot: Ashgate: 529–47.

Cullen, Lord (1990) *The Public Inquiry into the Piper Alpha Disaster (2 volumes), Cmnd 1310*, London: HMSO.

Cullen, Lord (2001) *The Ladbroke Grove Inquiry: Part 1 Report*, Norwich: HMSO.

Cutler, T. and James, P. (1996) 'Does Safety Pay? A Critical Account of the Health and Safety Executive Document: "The Costs of Accidents"', *Work, Employment & Society*, 10(4): 755–65.

Dalton, A. (2000) *Consensus Kills. Health and Safety Tripartism: Hazard to Workers' Health*, London: AJP Dalton.

Daniels, C. and Marlow, P. (2005) *Literature Review on the Reporting of Workplace Injury Trends. HSL/2005/36*, Buxton: Health and Safety Laboratory, www.hse.gov.uk/research/hsl_pdf/2005/hsl0536.pdf, downloaded December 2, 2006.

Davies, N.V. and Teasdale, P. (1994) *The Costs to the British Economy of Work Accidents and Work-Related Ill-Health*, London: HSE Books.

Davis, C. (2000) *Corporate Violence, Regulatory Agencies and the Management and Deflection of Censure*, unpublished doctoral thesis, University of Southampton.

Davis, C. (2004) *Making Companies Safe: What Works?* London: Centre for Corporate Accountability.

Dawson, S., Willman, P., Bamford, M. and Clinton, A. (1988) *Safety at Work: The Limits of Self-Regulation*, Cambridge: Cambridge University Press.

de Haan, W. (1990) *The Politics of Redress: Crime, Punishment and Penal Abolition*, London: Unwin Hyman.

Defert, D. (1991) '"Popular Life" and Insurance Technology', in G. Burchall, C. Gordon, P, Miller (eds), *Foucault Effect: Studies in Governmentality with Two Lectures by and an Interview with Michel Foucault*, Brighton: Harvester Wheatsheaf: 211–234.

Demaret, L. and Khalef, A. (2004) *Two million work deaths a year: carnage is preventable, ILO says*, www.hazards.org/wmd/ilobriefing2004.htm, downloaded December 6, 2006.

Department of Transport (1987) *The Merchant Shipping Act 1894, mv Herald of Free Enterprise, Report of Court No 8074 (Sheen Report)*, London: HMSO.

Doogan, K. (2001) 'Insecurity and Long-Term Employment', *Work, Employment & Society*, 15(3): 419–41.

Doran, N. (1996) 'From Embodied "Health" to Official "Accidents": Class, Codification, and British Factory Legislation, 1831–1844', *Social & Legal Studies*, 5(4): 333–57.

Dowie, M. (1977) 'How Ford Put Two Million Firetraps on Wheels', *Business and Society Review*, 23(Fall): 46–55.

Downes, D. and Morgan, R. (2002) 'The Skeletons in the Cupboard: The Politics of Law and Order at the Turn of the Millennium', in M. Maguire, R. Morgan, and R. Reiner (eds) *The Oxford Handbook of Criminology, Third Edition*, Oxford: Clarendon.

Doyal, L., with Pennell, I. (1979) *The Political Economy of Health*, London: Pluto.

Dunford, L. and Ridley, A. (1996) 'No Soul to Be Damned, No Body to Be Kicked: Responsibility, Blame and Corporate Punishment', *International Journal of the Sociology of Law*, 24(1): 1–19.

Equal Opportunities Commission (2005) *Facts About Men and Women in Britain*, at www.eoc.org.uk/pdf/facts_about_GB_2005.pdf, downloaded, December 2, 2006.

Etzioni, A. (1993) 'The US Sentencing Commission on Corporate Crime: A Critique', in G. Geis and P. Jesilow (eds) *White-Collar Crime*, Philadelphia: The Annals of the American Academy of Political and Social Science.

European Commission (2002) *European Social Statistics. Accidents at Work and Work-Related Health Problems. Data 1994–2000*, Luxemburg: European Commission.

Eurostat (2001) *European Statistics on Accidents at Work. Methodology*, Luxemburg: European Commission.

Everest, L. (1986) *Behind the Poison Cloud: Union Carbide's Bhopal Massacre*, New York: Banner Press.

Ewald, F. (1991) 'Insurance and Risk', in G. Burchall, C. Gordon and P. Miller (eds), *The Foucault Effect: Studies in Governmentality with Two Lectures by and an Interview with Michel Foucault*, Brighton: Harvester Wheatsheaf: 197–210.

Field, S. and Jorg, N. (1991) 'Corporate Liability and Manslaughter: Should we be Going Dutch?', *Criminal Law Review*, 156–71.

Fisse, B. (1971) 'The Use of Publicity as a Criminal Sanction Against Business', *Melbourne University Law Review*, 8: 107–50.

Fisse, B. (1990) 'Recent Developments in Corporate Criminal Law and Corporate Liability to Monetary Penalties', *University of New South Wales Law Journal*, 13: 1–19.

Fisse, B. and Braithwaite, J. (1983) *The Impact of Publicity on Corporate Offenders*, Albany, NY: State University of New York Press.
Fisse, B. and Braithwaite, J. (1993) *Corporations, Crime and Accountability*, Cambridge: Cambridge University Press.
Foucault, M. (1977) 'Prison Talk: An Interview with Michel Foucault', *Radical Philosophy*, 16: 10–15.
French, M. and Phillips, J. (2000) *Cheated Not Poisoned? Food Regulation in the United Kingdom, 1875–1938*, Manchester: Manchester University Press.
Friedrichs, D. (1998) 'Interview', in *Corporate Crime Reporter*, November 30th.

Galtung, J. (1981) *Violence and its Causes*, Paris: UNESCO.
Garland, D. (1994) 'Of Crimes and Criminals: The Development of Criminology in Britain', in M. Maguire, R. Morgan and R. Reiner (eds), *The Oxford Handbook of Criminology*, Oxford: Clarendon.
Geis, G. (1996) 'A Base on Balls for White Collar Criminals', in D. Shichor and D. Sechrest (eds) *Three Strikes and You're Out: Vengeance as Public Policy*, Thousand Oaks: Sage: 244–264.
Geis, G. and Dimento, J. (1995) 'Should We Prosecute Corporations and/or Individuals?', in F. Pearce and L. Snider (eds) *Corporate Crime: Contemporary Debates*, Toronto: University of Toronto Press: 72–86.
Geis, G. and Goff, C. (1983) 'Introduction', in E. Sutherland, *White Collar Crime: The Uncut Version*, London: Yale University Press: ix–xxxiii.
George, A. and Bennett, A. (2005) *Case Studies and Theory Development in the Social Sciences*, Cambridge, MA: The MIT Press.
Gill, S. and Law, D. (1988) *The Global Political Economy*, Hemel Hempstead: Harvester Wheatsheaf.
Glasbeek, H. (1989) 'Why Corporate Deviance is Not Treated as a Crime', in T. Caputo, M. Kennedy, C. Reasons and A. Brannigan (eds) *Law and Society: A Critical Perspective*, Toronto Harcourt Brace Jovanovich: 126–145.
Glasbeek, H. (2002) *Wealth by Stealth: Corporate Crime, Corporate Law and the Perversion of Democracy*, Toronto: Between the Lines.
Glasbeek, H. (2005) *Canada's Take on Corporate Killing: the Westray Bill*, London: Institute for Employment Rights.
Glasbeek, H. and Rowland, S. (1979) 'Are Injuring and Killing at Work Crimes?', *Osgoode Hall Law Journal*, 17: 506–94.
Glazebrook, P. (2002) 'A Better Way of Convicting Businesses of Avoidable Deaths and Injuries?', *Cambridge Law Journal*, 61(2): 405–22.
Gobert, J. and Punch, M. (2003) *Rethinking Corporate Crime*, London: Butterworths.

Goldblatt, D. (1997) 'At the Limits of Political Possibilities: The Cosmopolitan Democratic Project', *New Left Review*, 225: 140–50.

Goldman, L. (1994) 'Accident and Absolute Liability in Anthropology', in J. Gibbons (ed.), *Language and the Law*, London: Longman: 51–99.

Goodey, J. (2005) *Victims and Victimology. Research, Policy and Practice*, London: Longman.

Grabosky, P. (1997) 'Discussion Paper. Inside the Pyramid: Towards a Conceptual Framework for the Analysis of Regulatory Systems', *International Journal of the Sociology of Law*, 25(3): 195–201.

Gray, G. (2006) 'The Regulation of Corporate Violations. Punishment, Compliance, and the Blurring of Responsibility', *British Journal of Criminology*, 46(5): 875–92.

Gray, G. (2006a) 'Ticketing Health and Safety Offenders. A Socio-Legal Examination of Ticketing in High-Risk Firms', *Policy and Practice in Health and Safety*, 4(2): 77–97.

Grayson, J. and Goddard, C. (1976) '*Industrial Safety and the Trade Union Movement*, *Studies for Trades Unionists*, 1(4).

Green, G.S. (1990) *Occupational Crime*, Chicago: Nelson Hall.

Greenwood, P. (1983) 'Controlling the Crime Rate Through Imprisonment', in J.Q Wilson (ed.), *Crime and Public Policy*, San Francisco: Institute for Contemporary Studies: 251–69.

Grigg-Spall, I. and Ireland, P. (1992) *The Critical Lawyers' Handbook*, London: Pluto.

Gunningham, N. and Johnstone, R. (1999) *Regulating Workplace Safety: Systems and Sanctions*, Oxford: Oxford University Press.

Haines, F. (2005) 'Tracking the Regulation Debate', *The Australian and New Zealand Journal of Criminology*, 38(1): 141–7.

Hale, C. (2004) 'The Politics of Law and Order', in C. Hale, K. Hayward, A. Wahidin and E. Wincup (eds), *Criminology*, Oxford: Oxford University Press: 427–446.

Hall, A. and Johnstone, R. (2005) 'Exploring the Re-criminalisation of OHS Breaches in the Context of Industrial Death', *Flinders Journal of Law Reform*, 57(8): 57–92.

Hampton, P. (2005) *Reducing Administrative Burdens: Effective Inspection and Enforcement*, London: HM Treasury/HMSO.

Harpwood, V. (1993) *The Law of Tort*, London: Cavendish.

Harris, D. (1984) 'Claims for Damages', in D. Harris (ed.) *Compensation and Support for Illness and Injury*, Oxford: Clarendon Press.

Harrison, B. (1993) 'Are Accidents Gender-Neutral? The Case of Women's Work in Britain, 1880–1914', *Women's History Review*, 2(2): 253–75.

Hartley, H. (2001) *Explaining Sport and Leisure Disasters*, London: Cavendish.

Hawkins, K. (1984) *Environment and Enforcement: Regulation and the Social Definition of Pollution*, Clarendon Press: Oxford.
Hawkins, K. (1990) 'Compliance Strategy, Prosecution Policy and Aunt Sally: A Comment on Pearce and Tombs', *British Journal of Criminology*, 30(4): 444–66.
Hawkins, K. (1991) 'Enforcing Regulation: More of the Same from Pearce and Tombs', *British Journal of Criminology*, 31(4): 427–30.
Hawkins, K. (1996a) 'Compliance Strategy', in D. Galligan (ed.) *A Reader on Administrative Law*, Oxford: Oxford University Press: 299–325.
Hawkins, K. (1996b) 'Using Legal Discretion', in D. Galligan (ed.) *A Reader on Administrative Law*, Oxford: Oxford University Press: 247–273.
Hawkins, K. (ed) (1997) *The Human Face of Law*, Oxford: Clarendon Press.
Hawkins, K. (2002) *Law as a Last Resort: Prosecution Decision Making in a Regulatory Authority*, Oxford: Oxford University Press.
Hawkins, K. and Thomas, J. (eds) (1984) *Enforcing Regulation*, Boston: Kluwer.
Health and Safety Commission (1996) *Health and Safety Statistics, 1995/96*, Sudbury: HSE Books.
Health and Safety Commission (1997) *Health and Safety Statistics, 1996/97*, Sudbury: HSE Books.
Health and Safety Commission (1998) *National Picture of Health and Safety in Local Authority Enforced Industries 1998*, London: HSC/Government Statistical Service.
Health and Safety Commission (2000) *Health and Safety Statistics 1999/2000*, London: HSE Books.
Health and Safety Commission (2001) *Health and Safety Statistics 2000/01*, London: HSE Books.
Health and Safety Commission (2003) *Health and Safety Statistics Highlights 2002/03*, London: Health and Safety Commission/National Statistics, www.hse.gov.uk/statistics/overall/hssh0203.pdf, downloaded December 4, 2006.
Health and Safety Commission (2003a) *HSC Changing Patterns of Employment Programme*. Annual Report on Progress. HSC 03/91, http://www.hse.gov.uk/aboutus/hsc/meetings/2003/110203c91.pdf, downloaded December 2, 2006.
Health and Safety Commission (2004a) *Enforcement Policy Statement*, Sudbury: HSE Books.
Health and Safety Commission (2004b) *Strategy for Workplace Health and Safety in Great Britain to 2010 and Beyond*, Sudbury: HSE Books.
Health and Safety Commission (2005a) *Health and Safety Commission Annual Report and the Health and Safety Commission/Executive Accounts 2004/2005*, London: The Stationery Office.

Health and Safety Commission (2005b) *Directors Responsibilities for Improving Health and Safety Performance – Proposed Report to the Government, Health and Safety Commission paper HSC/05/90. Annex 1*, www.hse.gov.uk/aboutus/HSC/meetings/2005/061205/c90.pdf, downloaded, December 2, 2006.

Health and Safety Commission (2005c) *Directors Responsibilities for Improving Health and Safety Performance – Proposed Report to the Government, Health and Safety Commission paper HSC/05/90*, www.hse.gov.uk/aboutus/HSC/meetings/2005/061205/c90.pdf, downloaded, December 2, 2006.

Health and Safety Commission (2005d) *Minutes of the Health and Safety Commission*, HSC/06/M09.

Health and Safety Commission (2006a) *Health and Safety Statistics 2005/06*, London: National Statistics.

Health and Safety Commission (2006b) *Statistics of Fatal Injuries*, www.hse.gov.uk/statistics/overall/fatl0506.pdf, downloaded December 2, 2006.

Health and Safety Commission/Department of Environment, Transport and the Regions (2000) *Revitalising Health and Safety. Strategy Statement*, London: Department of Environment, Transport and the Regions.

Health and Safety Executive (1993) *The Costs of Accidents at Work*, London: HSE Books.

Health and Safety Executive (1996) *A Guide to the Reporting of Injuries, Diseases and Dangerous Occurrences Regulations 1995*, Sudbury: HSE Books.

Health and Safety Executive (1997) *Press Release E133:97 – July 28, 1997. Headline Workplace Health and Safety Statistics 1996/97.*

Health and Safety Executive (2000a) *Statistics of Workplace Fatalities and Injuries in Great Britain. International Comparisons 2000*, London: HSE.

Health and Safety Executive (2000b) *Key Messages from the LFS for Injury Risks: Gender and Age, Job Tenure and Part-Time Working*. Available at: www.hse.gov.uk/statistics/keyart.pdf, downloaded December 4, 2006.

Health and Safety Executive (2003a) *HSE Press Release: E193:03, October 3 2003*, www.hse.gov.uk/press/2003/e03193.htm.

Health and Safety Executive (2003b) *Statistics of Fatal Injures 2002/03*, London: Health and Safety Executive/National Statistics, available at: http://213.212.77.20/statistics/overall/fatl0203.pdf, downloaded December 4, 2006.

Health and Safety Executive (2003c) *Prosecuting Individuals*, Operational Circular OC 130/8, HSE internal policy document.

Health and Safety Executive (2004a) *HSE Publishes Investigation into Fatal Gas Explosion in Dundee*, Press Release, April 5.

Health and Safety Executive (2004b) *Health and Safety Statistics Highlights 2003/04*, Bootle: HSE Statistics Co-ordination Unit.

Health and Safety Executive (2004c) *Thirty Years On and Looking Forward: The Development and the Future of the Health and Safety System in Great Britain*, London: HSE Books, at www.hse.gov.uk/aboutus/reports/30years.pdf, downloaded December 2, 2006.

Health and Safety Executive (2005) *Health and Safety Offences and Penalties 2004–05. A Report by the Health and Safety Executive*, Sudbury: HSE Books.

Health and Safety Executive (2006) *Health and Safety Offences and Penalties 2003/04 – Key Statistics*, available online at http://www.hse.gov.uk/enforce/off03-04/statistics.htm, downloaded December 2, 2006.

Health and Safety Executive /Department for Transport (2003) *Driving at Work. Managing Work-Related Road Safety*, *INDG382*, Sudbury: HSE Books.

Herbert, I. and Nash, E. (2004) 'How Spain's insatiable appetite fuels a desperate pursuit of cockles in Morecambe', *The Independent on Sunday*, February 15, 2004.

Hertz, N. (2001) *The Silent Takeover: Global Capitalism and the Death of Democracy*, London: Arrow.

Hills, S. (1987) 'Preface', in S. Hills (ed.), *Corporate Violence. Injury and Death for Profit*, Totowa, NJ: Rowman and Littlefield: vii–viii.

Hills, S. (1987a) 'Epilogue: Corporate Violence and the Banality of Evil', in S. Hills (ed.), *Corporate Violence. Injury and Death for Profit,* Totowa, NJ: Rowman and Littlefield.187–206.

Hills, S. (ed) (1987b) *Corporate Violence. Injury and Death for Profit*, Totowa, NJ: Rowman and Littlefield:

Hillyard, P. (2006) 'Criminal Obsessions: Crime Isn't the Only Harm', *Criminal Justice Matters*, 62(Winter): 26–27, 46.

Hillyard, P. and Tombs, S. (2004) 'Beyond Criminology?', in P. Hillyard, C. Pantazis, S. Tombs and D. Gordon (eds), *Beyond Criminology? Taking Harm Seriously*, London: Pluto Press, 10–29.

Hillyard, P., Sim, J., Tombs, S. and Whyte, D. (2003) 'Leaving a "Stain Upon the Silence": Critical Criminology and the Politics of Dissent', *British Journal of Criminology*, 44(3): 369–90.

Hillyard, P., Pantazis, C., Tombs, S. and Gordon, D. (eds) (2004) *Beyond Criminology? Taking Harm Seriously*, London: Pluto Press.

Hilpern, K. (2004) 'Beware: danger at work', The *Guardian*, September 27, 2004.

von Hirsch, A. (1976/1996) 'Giving Criminals Their Just Deserts, Civil Liberties Review, no 3', in J. Muncie, E. McLaughlin and M. Langan (eds) *Criminological Perspectives: A Reader*, London: Sage: 315–324.

Hodge, N. (2002) *Scotland: Gas Monopoly Faces Killing Charge over Death of Family*, World Socialist Website, March 15, available at www.wsws.org/articles/2002/mar2002/tran-m15_prn.shtml, downloaded December 2, 2006.

Hopkins, A. (1994) 'Compliance with What? The Fundamental Regulatory Question', *British Journal of Criminology*, 34(4): 431–43.

Hughes, P.W. (1984) 'Cost Effective Health and Safety', *The Safety Practitioner*, May.

Hulsman, L. (1986) 'Critical Criminology and the Concept of Crime', in H. Bianchi and R. van Swaaningen (eds), *Abolitionism. Towards a Non-Repressive Approach to Crime*, Amsterdam: Free University Press: 25–41.

Hutter, B. (1988) *The Reasonable Arm of the Law? The Law Enforcement Procedures of Environmental Health Officers*, Oxford: Clarendon Press.

Hutter, B. (1997) *Compliance, Regulation and Environment*, Oxford: Clarendon Press.

Hutter, B. (2001) *Regulation and Risk: Occupational Health and Safety on the Railways*, Oxford: Oxford University Press.

Huws, U. (1987) *VDU Hazards Handbook*, London Hazards Centre.

International Labour Organisation (2005) *World Day for Safety and Health at Work 2005: A Background Paper*, Geneva: International Labour Office.

Ives, J. (ed) (1985) *The Export of Hazard. Transnational Corporations and Environmental Control Issues*, Boston: Routledge and KeganPaul.

James, P. and Walters, D. (2002) 'Worker Representation in Health and Safety: Options for Regulatory Reform', *Industrial Relations Journal*, 33(2): 141–156.

James, P. and Walters, D. (2005) *Regulating Health and Safety at Work: An Agenda for Change*, London: Institute of Employment Rights.

Jefferson, M. (2001) *Criminal Law*, London: Longman.

Jessop, B. (2002) *The Future of the Capitalist State*, Cambridge: Polity.

Johnstone, R. (1999) 'Putting the Regulated Back into Regulation', *Journal of Law and Society*, 26(3): 378–90.

Johnstone, R. (2003) *From Fiction to Fact - Rethinking OHS Enforcement, National Research Centre for OHS Regulation Working Paper 11*, at www.ohs.anu.edu.au/publications/pdf/wp%2011%20%20Johnstone.pdf, downloaded December 3, 2006.

Jolliffe, E. (2004) 'Comment. Pass the Safety Parcel', *The Safety and Health Practitioner*, May.

Jones, T. (1988) *Corporate Killings. Bhopals Will Happen*, London: Free Association Books.

Kalelkar, A. (1988) 'Investigation of Large Magnitude Incidents – Bhopal as a Case Study', in *Institute of Chemical Engineers Symposium Series No.110. Preventing Major Chemical and Related Process Accidents*, Rugby: IChemE: 553–575.

Kaye, T. (1992) 'Corporate Manslaughter: Who Pays the Ferryman?', in D. Feldman and F. Meisel (eds) *Corporate and Commercial Law: Modern Developments*, London: Lloyds of London.

Kelso, P. (2001) 'Student killed on first day at docks. Employer and manager deny charges of manslaughter', The *Guardian*, November 8.

Kramer, R. C. (1984) 'Corporate Criminality: The Development of an Idea', in E. Hochstedler (ed.), *Corporations as Criminals*, Beverly Hills: Sage Publications, 13–37.

Kramer, R. (1989) 'Criminologists and the Social Movement Against Corporate Crime', *Social Justice*, 16(2): 146–64.

Labour Research Department (1996) *Women's Health and Safety*, London: Labour Research Department.

Labour Research Department (2006) 'Survey Exposes FTSE Directors', Luxury Pension Packages, Press Release, 3 September.

Lacey, N. (ed) (1994) 'Introduction', *A Reader on Criminal Justice*: Oxford: Oxford University Press: 1–35.

Lacey, N. (1995) 'Contingency and Criminalisation', in I. Loveland (ed.), *Frontiers of Criminality*, London: Sweet and Maxwell: 1–28.

Lange, B. (1999) 'Compliance Construction in the Context of Environmental Regulation', *Social and Legal Studies*, 8(4): 549–67.

Lavalette, M. and Kennedy, J. (1996) *Solidarity on the Waterfront. The Liverpool Lock-Out of 1995/96*, Birkenhead: the Liver Press.

Lawrence, F. (2001) 'Running Up a Profit', The *Guardian*, May 17.

Lee, J. and Wrench, J. (1980) 'Accident-Prone Immigrants: An Assumption Challenged', *Sociology*, 14(4): 551–66.

Leopold, J.W. (1985) 'Workers Participation and Joint Union-Management Health and Safety Committees in the United Kingdom', in S. Bangara, R. Misiti and H. Wintersberger (eds), *Work and Health in the 1980s. Experiences of Direct Workers' Participation in Occupational Health*, Berlin: Wissenscahaftszentrum.

Lessin, N. and O'Neill, R. (2002) 'Bad Behaviour', *Health and Safety Bulletin*, 312, (October): 24–27.

Levi, M. (1997) 'Violent Crime', in M. Maguire, R. Morgan and R. Reiner (eds), *The Oxford Handbook of Criminology: Second Edition*, Oxford: Clarendon Press: 841–889.

Levi, M. (2000) 'Shaming and the Regulation of Fraud and Business "Misconduct": Some Preliminary Explorations', in S. Karstedt and K. Bussman (eds) *Social Dynamics of Crime and Control: New Theories for a World in Transition*, Oxford: Hart: 117–132.

Levi, M. (2002) 'Suite Justice or Suite Charity: Some Explorations of Shaming and Incapacitating Business Fraudsters', *Punishment and Society*, 4(2): 147–63.

Levitas, R. (1996) 'The Legacy of Rayner', in R. Levitas and W. Guy (eds), *Interpreting Official Statistics*, London: Routledge.

Lofquist, W. (1993) 'Organisational Probation and the US Sentencing Commission', in G. Geis and P. Jesilow (eds) *White Collar Crime*, Newbury Park, California: Sage: 157–69.

London Assembly Health and Public Services Committee (2005) *Building London, Saving Lives. Improving Health and Safety in Construction*, London: Greater London Authority.

Lynch, M. and Stretesky, P. (2003) 'The Meaning of Green: Contrasting Criminological Perspectives', *Theoretical Criminology*, 7(2): 217–38.

MacKenzie, G. (1699) *Law and Customes of Scotland in Matters Criminal*, Edinburgh: Andrew Anderson.

McMullan, J. and McClung, M. (2006) 'The Media, the Politics of Truth, and the Coverage of Corporate Violence: The Westray Disaster and the Public Inquiry', *Critical Criminology*, 14(1): 67–86.

Magistrates Association (2001) *Fining of Companies for Environmental and Health and Safety Offences: Magistrates Court Sentencing Guidelines, issued May 2001*, London: Magistrates Association.

Maguire, M. (1994) 'Crime Statistics, Patterns, and Trends: Changing Perceptions and their Implications', in M. Maguire, R. Morgan and R. Reiner (eds), *The Oxford Handbook of Criminology*, Oxford: Clarendon Press.

Maguire, M. (1997) 'Crime Statistics, Patterns and Trends: Changing Perceptions and Their Implications', in M. Maguire, R. Morgan and R. Reiner (eds), *The Oxford Handbook of Criminology: Second Edition*, Oxford: Clarendon Press.

Mahon, R. (1979) 'Regulatory Agencies: Captive Agents or Hegemonic Apparatuses?' *Studies in Political Economy*, 1(1): 162–200.

Marine Accident Investigation Branch (2005) *Marine Accident Investigation Branch Annual Report 2004*, Southampton: Marine Accident Investigation Branch.

Marx, K. (1954/1887) *Capital: Volume 1*, London: Lawrence and Wishart.

Marx, K. and Engels, F. (1890/1968) 'The Manifesto of the Communist Party', in *Karl Marx and Frederick Engels, Selected Works*, London: Lawrence and Wishart: 35–68.

Mayhew, C. and Quinlan, M. (1998) *The Effects of Outsourcing on Occupational Health and Safety: A Comparative Study of Factory-Based and Outworkers in the Australian TCF Industry*, Occupational Health and Safety Commission, Australian Government Publishing Service, Canberra.

Mayhew, C. Quinlan, M. and Ferris, R. (1997), 'The Effects of Subcontracting/Outsourcing on Occupational Health and Safety: Survey Evidence from Four Australian Industries', *Safety Science*, 25: 163–78.
Mathiesen, T. (1990) *Prison on Trial: Second Edition*, Winchester: Waterside Press.
Mathiesen, T. (2004) *Silently Silenced. Essays on the Creation of Acquiescence in Modern Society*, Winchester: Waterside Press.
McBarnet, D. (1988) 'Law, Policy and Legal Avoidance: Can Law Effectively Implement Egalitarian Policies?', *Journal of Law and Society*, 15(1): 113–21.
McBarnet, D. and Whelan, C. (1991) 'The Elusive Spirit of the Law: Formalism and the Struggle for Legal Control', *Modern Law Review*, 54(6): 848–73.
McCartney, J. (2005) 'Foreword' to *London Assembly Health and Public Services Committee, Building London, Saving Lives. Improving Health and Safety in Construction*, London: Greater London Authority.
Meier, R.F. and Short, J.F. Jnr. (1995) 'The Consequences of White-Collar Crime', in G. Geis, R.F. Meier and L. Salinger (eds), *White-Collar Crime. Classic and Contemporary Views. Third Edition*, New York: The Free Press: 80–104.
Michalowski, R. and Kramer, R. (1987) 'The Space Between Laws: The Problem of Corporate Crime in a Transnational Context', *Social Problems*, 34(1): 34–53.
Mill, J.S. (1859/1962) 'On Liberty', in M. Warnock (ed.) *Utilitarianism*, London: Fontana.
Monbiot, G. (2000) *Captive State: The Corporate Takeover of Britain*, London: MacMillan.
Moore, C. (1987) 'Taming the Giant Corporation? Some Cautionary Remarks on the Deterrability of Corporate Crime', *Crime and Delinquency*, 33(3): 379–402.
Moore, R. (1991) *The Price of Safety: The Market, Workers' Rights and the Law*, London: Institute of Employment Rights.
Moran, M. and Wright, M. (eds) (1991) *The Market and the State: Studies in Interdependence*, London: MacMillan.
Morehouse, W. and Subramaniam, M.A. (1986) *The Bhopal Tragedy*, New York: Council on International and Public Affairs.
Munro, W. (2006) 'Market Discipline', paper to the *Neo-liberal Scotland Conference*, University of Strathclyde, May 19–21.
Murphy, J.G. (1973/1995) 'Marxism and Retribution', in A. Duff, and D. Garland, (eds) *A Reader on Punishment*, Oxford University Press: Oxford.

Nelken, D. (1994) ‘White Collar Crime’, in M. Maguire, R. Morgan and R. Reiner (eds) *The Oxford Handbook of Criminology*, Oxford: Clarendon Press: 355–392.

Neocleous, M. (2003) *Imagining the State*, Maidenhead: Open University Press.

Nichols, T. (1986) *The British Worker Question*, London: RKP.

Nichols, T. (1989) ‘The Business Cycle and Industrial Injuries in British Manufacturing over a Quarter of a Century: Continuities in Industrial Injury Research’, *Sociological Review*, 37(3): 538–50.

Nichols, T. (1990) ‘Industrial Safety in Britain and the 1974 Health and Safety at Work Act: The Case of Manufacturing’, *International Journal of the Sociology of Law*, 18: 317–342.

Nichols, T. (1994) ‘Problems in Monitoring the Safety Performance of British Manufacturing at the End of the Twentieth Century’, *Sociological Review*, 37(3): 538–50.

Nichols, T. (1997) *The Sociology of Industrial Injury*, London: Mansell.

Nichols, T. and Armstrong, P. (1973) *Safety or Profit: Industrial Accidents and the Conventional Wisdom*, Bristol: Falling Wall Press.

Norrie, A. (2001) *Crime, Reason and History: A Critical Introduction to Criminal Law*, 2nd edn, London: Butterworths.

O‘Donnell, C. and White, L. (1998) *Invisible Hands. Child Employment in North Tyneside*. London: Low Pay Unit.

O’Donnell, C. and White, L. (1999) *Hidden Danger: Injuries to Children at Work in Britain*. London: Low Pay Unit.

O’Neill, R. (2006) ‘Come Clean’, *Hazards Magazine*, 95, July/Sept: 6–7.

Ogus, A. (1994) *Regulation: Legal Form and Economic Theory*, Oxford: Oxford University Press.

Osler, D. (2002) *Labour Party Plc. New Labour as a Party of Business*, Edinburgh: Mainstream.

Passas, N. (1990) ‘Anomie and Corporate Deviance’, *Contemporary Crises*, 14: 157–78.

Pate-Cornell, E. (1993) ‘Learning From the Piper Alpha Incident: A Post-Mortem Analysis of Technical and Organisational Factors, *Risk Analysis*, 13(2): 60–78.

Pearce, F. (1976) *Crimes of the Powerful: Marxism, Crime and Deviance*, London: Pluto.

Pearce, F. and Snider, L. (eds) (1995) *Corporate Crime: Contemporary debates*, Toronto: University of Toronto Press.

Pearce, F. and Tombs, S. (1990) ‘Ideology, Hegemony and Empiricism: Compliance Theories of Regulation’, *British Journal of Criminology*, 30(4): 423–43.

Pearce, F. and Tombs, S. (1991) 'Policing Corporate "Skid Rows": A Reply to Keith Hawkins', *British Journal of Criminology*, 31(4): 415–26.

Pearce, F. and Tombs, S. (1992) 'Realism and Corporate Crime', in R. Matthews and J. Young (eds) *Issues in Realist Criminology*, London: Sage: 70–101.

Pearce, F. and Tombs, S. (1993) 'US Capital Versus the Third World: Union Carbide and Bhopal', in F. Pearce and M. Woodiwiss (eds) *Global Connections: National and International Aspects of Crime and Crime Control*, London: MacMillan: 187–211.

Pearce, F. and Tombs, S. (1997) 'Hazards, Law and Class: Contextualising the Regulation of Corporate Crime', *Social & Legal Studies*, 6(1): 79–107.

Pearce, F. and Tombs, S. (1998) *Toxic Capitalism: Corporate Crime in the Chemical Industry*, Aldershot: Ashgate.

Peet, R. (2003) *Unholy Trinity: The IMF, The World Bank and WTO*, London: Zed.

Peltzman, S. (1976) 'Towards a More General Theory of Regulation', *The Journal of Law and Economics*, 19(2): 211–40.

Pemberton, S. (2004) 'A Theory of Moral Indifference: Understanding the Production of Harm by Capitalist Society', in P. Hillyard, C. Pantazis, S. Tombs and D. Gordon (eds), *Beyond Criminology? Taking Harm Seriously*, London: Pluto Press: 67–83.

Perrow, C. (1984) *Normal Accidents. Living with High-Risk technologies*, New York: Basic Books.

Pickvance, S. (2005) 'A Little Compensation', *Hazards*, 90, April/June: 18–19.

Pickvance, S. (2006) *Preventing Ill-Health Caused by Work: What Progress?* Paper presented at Health and Safety – revitalised or reversed?, London: Institute of Employment Rights, January 18.

Piquero, N., Exum, M. and Simpson, S. (2005) 'Integrating the Desire-for-Control and Rational Choice in a Corporate Crime Context', *Justice Quarterly*, 22(2): 252–80.

Polanyi, K. (1944/1957) *The Great Transformation: The Political and Economic Origins of Our Time*, Boston: Beacon.

Punch, M. (1996) *Dirty Business: Explaining Corporate Misconduct*, London: Sage.

Punch, M. (2000) 'Suite Violence: Why Managers Murder and Corporations Kill', *Crime, Law and Social Change*, 33: 243–80.

Reilly, B., Pace, P. and Hall, P. (1995) 'Unions, Safety Committees and Workplace Injuries', *British Journal of Industrial Relations*, 33(2): 273–88.

Reiman, J. (1998) *The Rich Get Richer and the Poor Get Prison*. Fifth Edition, Boston: Allyn & Bacon.

Reiman, J.H. (1979) *The Rich Get Richer and the Poor Get Prison. Ideology, Class, and Criminal Justice*, New York: John Wiley & Sons.

Ridley, A. and Dunford, L. (1997) 'Corporate Killing – Legislating for Unlawful Death?' *Industrial Law Journal*, 26(2): 99–113.

Robens, Lord (1972) *Safety and Health at Work, Report of the Committee 1970–72. Cmnd. 5034*, London: HMSO.

Robertson, H. (2004) 'Bad Language', *Hazards*, 86, April/June: 6–7.

Royal Society for the Prevention of Accidents (RoSPA) (1998) *Managing Occupational Road Risk*, Birmingham: RoSPA.

Rusche, G. (1933/1980) 'Labour Market and Penal Sanction: Thoughts on the Sociology of Criminal Justice', in T. Platt and P. Takagi (eds) *Punishment and Penal Discipline*, Berkley California: Crime and Justice Associates: 10–17.

Rusche, G. and Kirchheimer, O. (1939/2003) *Punishment and Social Structure*, Transaction: New Jersey.

Ryan, W. (1971) *Blaming the Victim*, New York: Pantheon.

Ryggvik, H. (2000) 'Offshore Safety Regulations in Norway: From Model to Systems in Erosion', *New Solutions*, 10(1–2): 67–116.

Salmi, J. (1993) *Violence and Democratic Society. New Approaches to Human Rights*, London: Zed Books.

Salmi, J. (2004) 'Violence in Democratic Societies: Towards an Analytic Framework', in P. Hillyard, C. Pantazis, S. Tombs and D. Gordon (eds), *Beyond Criminology? Taking Harm Seriously*, London: Pluto Press: 55–66.

Sanders, A. (1985) 'Class Bias in Prosecutions', *Howard Journal*, 24(3): 176–99.

Sass, R. and Crook, G. (1981) 'Accident-Proneness: Science or Non-Science?', *International Journal of Health Services*, 11(2): 175–90.

Scheppele, K.L. (1991) 'Law Without Accidents', in P. Bordieu and J.S. Coleman (eds), *Social Theory for a Changing Society*, Boulder: Westview: 267–93.

Select Committee on Environment, Transport and Regional Affairs, Fourth Report, *The Work of the Health and Safety Executive* (2000) HC31, London: The Stationery Office.

Shapiro, S.P. (1983) 'The New Moral Entrepreneurs: Corporate Crime Crusaders', *Contemporary Sociology*, 12: 304–7.

Schrager, L.S. and Short, J.F. (1977) 'Towards a Sociolgy of Organisational Crime', *Social Problems*, 25: 407–19.

Shrivastava, P. (1992) *Bhopal: Anatomy of a Crisis*, London: Paul Chapman Publishing.

Sklair, L. (2001) *The Transnational Capitalist Class*, Oxford: Blackwell.

Slapper, G. (1993) 'Corporate Manslaughter: An Examination of the Determinants of Prosecutorial Policy', *Social and Legal Studies*, 2(4): 423–443.

Slapper, G. (2000) *Blood in the Bank. Social and Legal Aspects of Death at Work*, Aldershot: Ashgate.

Slapper, G. and Tombs, S. (1999) *Corporate Crime*, London: Longman.

Smith, A., Wadsworth, E., Shaw, C., Stansfeld, S., Bhui, K. and Dhillon, K. (2005) *Ethnicity, Work Characteristics, Stress and Health. Health and Safety Executive Research Report 308*. London: HSE Books.

Smith, C. and Allen, J. (2004) *Violent Crime in England and Wales. Home Office Online Report 18/04*, www.homeoffice.gov.uk/rds/pdfs04/rdsolr1804.pdf, downloaded December 2, 2006.

Smith, D. and Tombs, S. (1995) 'Beyond Self-Regulation: Towards a Critique of Self-Regulation as a Control Strategy for Hazardous Activities', *Journal of Management Studies*, 35(5): 619–36.

Smith, J. and Hogan, B. (2002) *Criminal Law*, 10th edn, London: Lexis Nexis Butterworths Tolley.

Smith, P.M. and Mustard, C.A. (2004) 'Examining the Associations Between Physical Work Demands and Work Injury Rates Between Men and Women in Ontario, 1990–2000', *Occupational and Environmental Medicine*, 61: 750–56.

Snider, L. (1991) 'The Regulatory Dance: Understanding Reform Processes in Corporate Crime', *International Journal of Sociology of Law*, 19(2): 209–37.

Snider, L. (1993) *Bad Business: Corporate Crime in Canada*, Toronto: Nelson.

Snider, L. (2000) 'The Sociology of Corporate Crime: An Obituary (or: Whose Knowledge Claims Have Legs?)', *Theoretical Criminology*, 4(2): 169–206.

Snider, L. (2003) 'Captured by Neo-liberalism: Regulation and Risk in Walkerton, Ontario', in *Risk Management*, 5(2): 17–27.

Soros, G. (1998) *The Crisis of Global Capitalism: Open Society Endangered*, London: Little Brown and Company.

Southern and Eastern Regional Trades Union Congress (2005) *Evidence from Southern and Eastern Regional Trades Union Congress to the Greater London Assembly Scrutiny Inquiry. Health and Safety of Construction Workers in London*, http://www.tuc.org.uk/h_and_s/tuc-9752-f0.cfm?regional=7, downloaded December 2, 2006.

Spencer, R. (2002) *Corporate Law and Structures: Exposing the Roots of the Problem*, Oxford: Corporate Watch.

Start, R.D., Delargy-Aziz, Y., Dorries, C.P., Silcocks, P.B. and Cotton, D.W.K. (1993) 'Clinicians and the Coronial System: Ability of Clinicians to Recognise Reportable Deaths', *British Medical Journal*, 306(April 17th): 1038–41.

Start, R.D., Usherwood, T.P., Carter, N., Dorries, C.P. and Cotton, D.W.K. (1995) 'General Practitioners' Knowledge of When to Refer Deaths to a Coroner', *British Journal of General Practice*, 45: 191–3.

Stevens, G. (1992) 'Workplace Injury: a View from HSE's Trailer to the 1990 Labour Force Survey', *Employment Gazette*, December, 621–38.
Stigler, G. (1971) 'The Theory of Economic Regulation', *Bell Journal of Economics and Managerial Science*, 2(3): 3–21.
Stolberg, S. and Harris, G. (2003) 'Industry fights to put imprint on drug bill', *New York Times*, September 5.
Sullivan, B. (2001) 'Corporate Killing – Some Government Proposals', *Criminal Law Review*, January, 31–39.
Sumner, C. (ed) (1990) *Censure, Politics and Criminal Justice*, Milton Keynes: Open University Press.
Sutherland, E. (1983) *White Collar Crime, The Uncut Version*, New Haven: Yale University Press.
Sutherland, E. (1940) 'White-Collar Criminality', reprinted in G. Geis, R.F. Meier and L.M. Salinger (eds) (1995) *White-Collar Crime. Classic and Contemporary Views*, New York: The Free Press, 29–38.
Sutherland, E. (1945) 'Is "White-Collar Crime" Crime?', *American Sociological Review*, 10(2): 132–39.
Sutherland, E. (1949) *White-Collar Crime*, New York: Holt Reinhart and Winston.
Sutherland, E. (1983) *White Collar Crime. The Uncut Version*, New Haven: Yale University Press.
Szockyi, E. and Fox, J.G. (eds) (1996) *Corporate Victimisation of Women*, Boston, MA: Northeastern University Press.
Szockyi, E. and Frank, N. (1996) 'Introduction', in E. Szockyi and J.G. Fox (eds), *Corporate Victimisation of Women*, Boston, MA: Northeastern University Press, 3–32.

Takala, J. (2002) Introductory Report: Decent Work – Safe Work, paper presented at *XVIth World Congress on Safety and Health at Work*, Vienna, May 26–31.
Tappan, P. (1947) 'Who Is the Criminal?', *American Sociological Review*, 12(1): 96–102.
Taylor, I., Walton, P. and Young, J. (1973) *The New Criminology*, London: RKP.
Thompson, E.P. (1975) *Whigs and Hunters: The Origin of the Black Act*, Harmondsworth: Penguin.
Tombs, S. (1990) 'Industrial Injuries in British Manufacturing', *Sociological Review*, 38(2): 324–343.
Tombs, S. (1991) 'Injury and Ill-Health in the Chemical Industry: De-centring the Accident-Prone Victim', *Industrial Crisis Quarterly*, 5 (January): 59–75.

Tombs, S. (1995) 'Law, Resistance and Reform: "Regulating" Safety Crimes in the UK', *Social & Legal Studies*, 4(3): 343–65.

Tombs, S. (1996) 'Injury, Death and the Deregulation Fetish: The Politics of Occupational Safety Regulation in UK Manufacturing Industries', *International Journal of Health Services*, 26(2): 309–26.

Tombs, S. (1999a) 'Death and Work in Britain', *Sociological Review*, 47(2): 345–67.

Tombs, S. (1999b) 'Health and Safety Crimes: (In)visibility and the Problems of "knowing"', in P. Davies, P. Francis. and V. Jupp (eds), *Invisible Crimes: Their Victims and their Regulation*, London: Macmillan: 77–104.

Tombs, S. (2000) 'Official Statistics and Hidden Crimes: Researching Health and Safety Crimes', in V. Jupp, P. Davies and P. Francis (eds) *The Practice of Criminological Research*, London: Sage: 64–81.

Tombs, S. (2001) 'Thinking About "White-Collar" Crime', in S-Å. Lindgren (ed.), *White-Collar Crime Research. Old Views and Future Potentials. Lectures and Papers from a Scandinavian Seminar* (BRÅ-Rapport 2001:1). Stockholm: Brottsförebyggande råde/Fritzes: 13–34.

Tombs, S. (2002) 'Understanding Regulation?', *Social and Legal Studies*, 11(1): 113–33.

Tombs, S. (2004) 'Workplace Injury and Death: Social Harm and the Illusion of Law', in P. Hillyard *et al.* (eds), *Beyond Criminology? Taking Harm Seriously*, London: Pluto Press: 156–77.

Tombs, S. (2005) 'Commentary. The Conventionalisation of Early Factory Crime', *Policy and Practice in Health and Safety*, 3(2): 103–5.

Tombs, S. (2006) 'Violence, Safety Crimes and Criminology', *British Journal of Criminology*, Advance Access published on December, doi: 10,1093/bjc/az1095.

Tombs, S. and Whyte, D. (1998) 'Capital Fights Back: Risk, Regulation and Profit in the UK Offshore Oil Industry', *Studies in Political Economy*, 57(September): 73–101.

Tombs, S. and Whyte, D. (2001) 'Media Reporting of Crime: Defining Corporate Crime Out of Existence?', *Criminal Justice Matters*, 43(Spring): 22–3.

Tombs, S. and Whyte, D. (2003a) 'Two Steps Forward, One Step Back: Towards Accountability for Workplace Deaths?', *Policy and Practice in Health and Safety*, 1(1): 9–30.

Tombs, S. and Whyte, D. (2003b) 'Corporations Beyond the Law? Regulation, Risk and Corporate Crime in a Globalised Era', *Risk Management*, 5(2): 9–16.

Tombs, S. and Whyte, D. (2006) 'Corporate Crime', in E. McLaughlin and J. Muncie (eds), *The Sage Dictionary of Criminology. Second Edition*, London: Sage: 74–76.

Trades Union Congress (2002) *No More 'Men Only' Health and Safety*, available online at: www.tuc.org.uk/h_and_s/tuc-4838-f0.cfm, downloaded December 2, 2006.

Trades Union Congress (2003a) *Migrant Workers – Overworked, Underpaid and Over Here*, Press Release, 10th July.

Trades Union Congress (2003b) *Risks*, 96, March 8, www.tuc.org.uk/h_and_s/tuc-6365-f0.cfm#f1, downloaded December 2, 2006.

Trades Union Congress (2003c) 'Employers are not meeting the costs of workplace injuries and illness', *TUC Press Release*, March 18.

Trades Union Congress (2005a) *Risks*, 201, April 9, www.tuc.org.uk/h_and_s/tuc-9652-f0.cfm#o1, downloaded December 2, 2006.

Trades Union Congress (2005b) *Risks*, 233, November 19, www.tuc.org.uk/h_and_s/tuc-11021-f0.cfm#tuc-11021-15, downloaded December 2, 2006.

Trades Union Congress (2005c) *TUC Submission, Women and Work Commission*, www.tuc.org.uk/equality/tuc-9391-f0.pdf, downloaded December 2, 2006.

Trades Union Congress (2005d) *Risks*, 212, June 25, www.tuc.org.uk/h_and_s/tuc-10093-f0.cfm#o3, downloaded December 2, 2006.

Trades Union Congress (2006) *HSE Funding, a case for more resources*, briefing available online at: http://www.tuc.org.uk/h_and_s/tuc-12542-f0.cfm, downloaded November 25, 2006.

Tucker, E. (1990) *Administering Danger in the Workplace. The Law and Politics of Occupational Health and Safety Regulation in Ontario 1950–1914*, Toronto: University of Toronto Press.

Tucker, E. (ed) (2006) *Working Disasters: The Politics of Recognition and Response*, New York: Baywood Press.

Tweedale, G. (2000) *Magic Mineral to Killer Dust: Turner and Newall and the Asbestos Hazard*, Oxford: Oxford University Press.

UK National Work Stress Network (2006) *Information Pack 2006*, Norwich: UK National Work Stress Network.

Underhill, G. (1994) 'Introduction: Conceptualising the Changing Global Order' in R. Stubbs and G. Underhill (eds) *Political Economy and the Changing Global Order*, New York: Palgrave MacMillan, 17–44.

Unison/Centre for Corporate Accountability (CCA) (2002) *Safety Last? The Under-Enforcement of Health and Safety Law. Full Report*, London: Unison/Centre for Corporate Accountability.

Vaughan, D. (1992) 'The Macro-Micro Connection in White-Collar Crime Theory', in K. Schlegel and D. Weisburd (eds) *White-Collar Crime Reconsidered*, Boston, Mass: Northeastern Universtiy Press: 124–45.

Vaughan, D. (1996) *The Challenger Launch Decision. Risky Technology, Culture, and Deviance at NASA*, Chicago: Chicago University Press.
Virta, E. (1999) 'A Thief is a Criminal Who has Not Had Enough Time to Start a Company', in A. Laitinen and V. Olgiati (eds), *Crime-Risk-Security*, Turku: University of Turku, 91–128.

Wallach, L. and Sforza, M. (1999) *The WTO: Five Years of Reasons to Resist Corporate Globalisation*, New York: Seven Stories Press.
Walters, D., Nichols, T., Conner, J., Tasiran, A. and Cam, S. (2005) *The Role and Effectiveness of Safety Representatives in Influencing Workplace Health and Safety, HSE Research Report 363*, London: HSE Books.
Walters, D.R. (1987) 'Health and Safety and Trade Union Workplace Organization: A Case Study in the Printing Industry', *Industrial Relations Journal*, 18(1): 40–9.
Walters, R. (2003) *Deviant Knowledge. Criminology, Politics and Policy*, Cullompton: Willan.
Walters, R. (2006) 'Crime, Bio-Agriculture and the Exploitation of Hunger', *British Journal of Criminology*, 46(1): 26–45.
Watterson, A. (1988) *Industrial Relations and Health and Safety at Work in Post-War Britain: A Study of Conflict and Control in the Workplace*, unpublished PhD thesis, University of Bristol.
Watterson, A. (1991) 'Occupational Health in the UK Gas Industry: A Study of Employer, Worker and Medical Knowledge and Action on Health Hazards in the Late 19th and Early 20th Centuries', paper presented to the *Annual Conference of the British Sociological Association*, Manchester, March.
Weisburd, D. and Waring, E. (2001) *White-Collar Crime and Criminal Career*, Cambridge: Cambridge University Press.
Wells, C. (1993) *Corporations and Criminal Responsibility*, Oxford: Clarendon.
Wells, C. (1995) 'Cry in the Dark: Corporate Manslaughter and Cultural Meaning', in I. Loveland (ed.), *Frontiers of Criminality*, London: Sweet & Maxwell: 109–125.
Wells, C. (2001) *Corporations and Criminal Responsibility, Second Edition*, Oxford: Clarendon.
White, R. (2003) 'Environmental Issues and the Criminological Imagination', *Theoretical Criminology*, 7(4): 483–506.
Whyte, D. (1999a) *Power, Ideology and the Regulation of Safety in the Post-Piper Alpha Offshore Oil Industry*, unpublished PhD thesis, Liverpool John Moores University.
Whyte, D. (1999b) 'Rail Crash Inquiry: A Public Stitch Up?', *Socio-Legal Newsletter*, no. 29, November.

Whyte, D. (2004) 'Regulation and Corporate Crime', in J. Muncie and D. Wilson (eds), *Student Handbook of Criminal Justice and Criminology*, London: Cavendish: 133–152.

Whyte, D. (2006) 'Regulating Safety, Regulating Profit: Cost Cutting, Injury and Death in the North Sea after Piper Alpha', in E. Tucker (ed.) *Working Disasters: The Politics of Recognition and Response*, New York: Baywood Press: 181–206.

Wilson, J. (1980) 'The Politics of Regulation', in J. Wilson (ed.) *The Politics of Regulation*, New York: Basic Books: 357–94.

Woolf, A. (1973) 'Robens – The Wrong Approach', *Industrial Law Journal*, 2(1): 88–95.

Woolfson, C. (2005) 'Un-Social Europe', *Transitions Online*, June 10, 2005, www.tol.cz/look/TOL/article.tpl?IdLanguage=1&IdPublication=4&NrIssue=119&NrSection=4&NrArticle=14149&tpid=6, downloaded December 2, 2006.

Woolfson, C. and Beck, M. (1997) *From Self-Regulation to Deregulation: The Politics of Health and Safety in Britain*, Mimeo, Universities of Glasgow and St Andrews.

Woolfson, C., Foster, J. and Beck, M. (1996) *Paying for the Piper: Capital and Labour in Britain's Offshore Oil Industry*, London: Mansell.

Work Hazards Group (1988) *Death at Work*, London: Workers' Educational Association.

Wrench, J. (1996) 'Hazardous Work: Ethnicity, Gender and Resistance', paper presented at the British Sociological Association Annual Conference, University of Reading, April 1–4.

Wrench, J. and Lee, J. (1982) 'Piecework and Industrial Accidents: Two Contemporary Case Studies', *Sociology*, 16(4): 512–25.

Index